Modern Weights and Measures Regulation in the United States

In this book, Craig A. Leisy provides a concise history of weights and measures regulation in the United States from the early 20th century up to the present day.

Written for academic and professional readers, Leisy describes basic terms and concepts, the origins and history of weights and measures laws, weights and measures regulation, the economics of regulation, key enforcement cases, landmark legal decisions, the effects of public policy and a forecast on the future of weights and measures regulation. He also discusses the impact of weights and measures regulation on both producers (sellers) and consumers (buyers) in the marketplace. The book also features a new 2019 survey of state weights and measures regulatory programs, an introduction to the economics of weights and measures regulation, a case study of the municipal weights and measures regulatory program in Seattle, Washington, details of a major gasoline dispenser fraud case in Los Angeles County and landmark legal cases related to net contents of packaged goods.

Modern Weights and Measures Regulation in the United States is the only book on this subject from the perspective of a former long-time weights and measures regulatory official.

Craig A. Leisy served as the Manager of the Consumer Affairs Unit for the City of Seattle where he was responsible for regulation of weights and measures from 1996 to 2017. He was a member of the Western Weights and Measures Association (WWMA) and the National Conference on Weights and Measures (NCWM). He was asked by the NCWM to lead the first survey of state weights and measures regulatory programs in 2002 and he conducted a follow-up survey in 2019 for this book. He graduated from the U.S. Coast Guard Academy in New London, Connecticut with a Bachelor of Science degree. He also holds a Master of Business Administration (MBA) degree from the University of Puget Sound and a Master of Marine Affairs (MMA) degree from the University of Washington. He taught economics and management for three years at the U.S. Coast Guard Academy. His previous book, *Transportation Network Companies and Taxis: The Case of Seattle*, was published by Routledge in 2019.

Routledge Research in Public Administration and Public Policy

Affirmative Action in Malaysia and South Africa
Preference for Parity
Hwok-Aun Lee

Critical Perspectives on Public Systems Management in India
Through the Lens of District Administration
Amar K J R Nayak and Ram Kumar Kakani

Policy Making and Southern Distinctiveness
John C. Morris, Martin K. Mayer, Robert C. Kenter & R. Bruce Anderson

The Effects of Technology and Institutions on E-Participation
A Cross-National Analysis
Pragati Rawat & John C. Morris

A Guidebook for City and County Managers
Meeting Today's Challenges
James M. Bourey

Modern Weights and Measures Regulation in the United States
A Brief History
Craig A. Leisy

COVID-19 Pandemic, Public Policy and Institutions in India
Issues of Labour, Income, and Human Development
Edited by Indranil De, Soumyadip Chattopadhyay, Hippu Salk Kristle Nathan, and Kingshuk Sarkar

For more information about this series, please visit: www.routledge.com/Routledge-Research-in-Public-Administration-and-Public-Policy/book-series/RRPAPP

Modern Weights and Measures Regulation in the United States

A Brief History

Craig A. Leisy

NEW YORK AND LONDON

First published 2022
by Routledge
605 Third Avenue, New York, NY 10158

and by Routledge
2 Park Square, Milton Park, Abingdon, Oxon, OX14 4RN

Routledge is an imprint of the Taylor & Francis Group, an informa business

Library of Congress Cataloging-in-Publication Data
A catalog record for this title has been requested

ISBN: 978-1-032-20460-4 (hbk)
ISBN: 978-1-032-20465-9 (pbk)
ISBN: 978-1-003-26366-1 (ebk)

DOI: 10.4324/9781003263661

Typeset in Times New Roman
by Newgen Publishing UK

This book is dedicated to my coworkers at the Consumer Affairs Unit for the City of Seattle while I served as the manager during 1996–2015. The success of the weights and measures regulatory program was entirely due to their hard work. I was just smart enough to get out of their way.

Contents

Illustrations

Figures

Photos

Tables

Preface

In 1996, when I was hired to manage the Consumer Affairs Unit for the City of Seattle, I had no background in weights and measures but I had considerable regulatory experience. I had been a commercial vessel safety inspector in the U. S. Coast Guard during my first career (1972–1995). As a result, I was very familiar with the basic work of a regulator. Midway in that career, I was selected for a three-year out-of-specialty assignment as an assistant professor in the Department of Economics and Management at the U.S. Coast Guard Academy in New London, Connecticut (1981–1984). I taught survey courses on economics and management as well as upper-level courses on transportation economics and transportation policy. Previously, I had earned a Master of Business Administration degree from the University of Puget Sound. Later, I earned a Master of Marine Affairs degree from the University of Washington in Seattle with an emphasis on ports and marine transportation. By the time I began my second career at the City of Seattle (1996–2017), I was already familiar with managing regulatory programs and the economics of regulation.

No one applies for a job in weights and measures regulation with relevant experience on their resume. There are fewer than two thousand weights and measures inspectors spread out all over the United States and its territories. Most manager positions are filled from within the state and local government regulatory programs so the candidates are at least familiar with the *nuts and bolts* of field inspector work. I was brought in from the outside so everything was new to me.

I looked but could not find a book on weights and measures regulation to help me learn my new job in Seattle. I thought that maybe someday I would write one based on what I learned. I began to collect information on all aspects of that work. Much of it eventually found its way into this book. There has always been a need for a brief introduction to weights and measures regulation but the only books available were the National Institute of Standards and Technology (NIST) Handbook series on technical standards, test procedures and model law. In 2011, NIST published Handbook 155 *Weights and Measures Program Requirements: A Handbook for the Weights and Measures Administrator*. This volume was a good start. My intent was to build on the foundations of that publication but broaden the scope and depth to reach a wider audience including federal, state, county and city weights and measures regulators; elected lawmakers and their staffs at all levels of government, private weighing and measuring device service companies and regulated industry – manufacturing, wholesale and retail businesses impacted by weights and measures laws. The book should also be of great value for academic researchers or instructors studying the impact of regulatory programs on the economy.

Introduction

Few people know the meaning of the term, '*weights and measures*,' and fewer still can explain what weights and measures officials do or even why they do it. This is ironic because weights and measures laws are applicable to nearly all goods, and even some services, that are sold in the United States and virtually all other developed economies in the world. According to transportation economists, almost everything that we consume is transported to stores on trucks. It is not an overstatement to say that almost everything in those trucks is subject to one or more weights and measures laws.

What is 'Weights and Measures'?

So, you may ask, "What is the meaning of *weights and measures*?" or "Why haven't I heard about weights and measures before?" Well, you have probably seen a weights and measures official in a grocery store or at a gas station and did not know it. Weights and measure officials that work in the 'field,' at retail businesses and production facilities, are usually referred to as '*inspectors*' because they are conducting inspections to verify compliance with weights and measures laws. Most inspectors are dressed like everyone else except that inspectors, at gas stations, warehouses and other *dirty* locations, often wear coveralls over their street clothes. Inspectors are often present when consumers are not. Inspectors attempt to minimize inconvenience to consumers, and store staff, by avoiding peak business hours. Peak hours are generally the morning and evening rush hours and at midday. Often, the only evidence that inspectors are present is a government vehicle, usually a pickup truck or a small truck with a service box, parked nearby. All official vehicles have required markings on their doors to identify them. Inspectors don't schedule inspections in advance so that they can observe the regular work practices of store employees and inspect the stores as

DOI: 10.4324/9781003263661-1

they are. For example, they observe checkout clerks at the '*point-of-sale*' to verify that they take proper '*tare*' (i.e., deduct the weight of any packaging) so that consumers are only charged for '*net weight*' or the weight of the product alone. The terms used here - '*point-of-sale*,' '*tare*,' and '*net weight*' – have precise technical-legal meanings and are defined in the Glossary at the conclusion of this book.

What Exactly Do Weights and Measures Inspectors Do?

You may also ask, "What exactly do weights and measures inspectors do?" A substantial portion of their activities consists of inspecting and testing commercial '*weighing and measuring devices*.' Weighing and measuring devices are used to weigh goods sold from '*bulk*' across a '*scale*' (e.g., solids) or to measure goods sold through a '*meter*' (e.g., liquids). Inspectors verify the accuracy of scales and meters using calibrated test weights or test measures. Weights and measures inspectors conduct many other inspection activities that apply specifically to goods sold in '*packaged form*.' Checking the '*net contents of packaged goods*' involves verifying that the quantity of product contained in packages, selected at random, is at least equal to the '*net weight*' (solids) or '*net contents*' (liquids) displayed on the package labels. Other inspection activities applicable to packaged goods are known as '*unit pricing*,' '*method of sale*' and '*price verification*.' These will all be discussed in detail in Chapter 1 and subsequent chapters as well as the Glossary. '*Price verification*' is the process of verifying that the '*item price*' on the product or shelf tag, and the lowest advertised price, are the same as the price on the cash register display (from the store computer) when the bar code is scanned. Virtually all retail stores, from small convenience stores to big retail box stores, scan bar codes of items at checkout including apparel, food, office supplies or products for sale in hardware stores. As a result, the majority of consumer goods counted in the nation's Gross Domestic Product (GDP) are subject to weights and measures price verification inspections.

This book is a brief history of weights and measures regulation in the United States. The case history of the weights and measures regulatory program in Seattle, Washington traces its evolution from the early days of the twentieth century up to the modern day. Several special features of this book are summarized here:

- First, a new 2019 survey of state weights and measures regulation in the United States was conducted by the author and included in Chapter 1, "Modern Weights and Measures Regulation in the

United States." The author led the first survey conducted in 2002 on behalf of the National Conference on Weights and Measures. The 2019 survey results are compared with the results of the 2002 survey to examine trends.

- Second, an introduction to the economics of weights and measures regulation, an often ignored but vital subject, is presented as Chapter 2, "The Economics of Weights and Measures Regulation in the United States." Weights and measures regulatory program budgets are under a lot of pressure so the economic costs and benefits of regulation are under increasing scrutiny.
- Third, a case history of a municipal weights and measures regulatory program appears as Chapter 3, "Case History: Weights and Measures Regulation in Seattle, Washington." This extended example helps illustrate how regulatory programs evolved and how they operate.
- Fourth, a detailed study of the important 1997 gas fraud case in Los Angeles County is included in Chapter 5, "Enforcement Issues." The study insights rely heavily on an extensive interview with the lead weights and measures investigator for Los Angeles County. It illustrates many aspects of computer fraud. And,
- Fifth, some of the most important legal cases dealing with the key issue of '*net contents*' are analyzed with reference to their influence on modern weights and measures regulation. These appear in Chapter 6 "Landmark Legal Cases."

It is hoped that this book on weights and measures regulation will provide professionals (regulatory officials, elected officials and their staffs, and representatives from the regulated industry) and academic scholars with the background needed for a basic understanding of this specialized field of regulation. This is the *only* book in print on modern weights and measures regulation in the United States and it was written by a former long-time weights and measures regulatory program manager with insights based on more than 20 years of experience.

1 Modern Weights and Measures Regulation in the United States

Weights and measures regulation in the 21st century is, in many respects, very similar to the 20th century except that the marketplace reflects changing consumer tastes and advances in technology that have, in turn, prompted corresponding changes to inspection types and practices. Weights and measures regulation is still conducted by state and local government agencies but there has been considerable consolidation as many municipalities and counties have turned over their programs to states due to budgetary difficulties.

First Survey of State Weights and Measures Programs [2002]

In 2003, the Chairman of the National Conference on Weights and Measures (NCWM) appointed the author of this book to lead a working group of representatives from all four regional associations comprising the NCWM (e.g., Western Weights and Measures Association) to conduct a first-ever national survey of state weights and measures regulatory programs. All 50 state weights and measures regulatory programs were sent a questionnaire about 2002 budgets, staffing, inspection workloads, device registration and inspection fees, and inspection procedures and policies. The purpose of the survey was to learn the fundamentals about state weights and measures regulatory programs in all states that comprise the NCWM.

> The purpose of the survey was to establish baseline information on the types of inspection statistics collected by state weights and measures programs. This is the first phase of a larger project intended to promote uniform data collection and to measure the value of weights and measures work nationwide.[1]

DOI: 10.4324/9781003263661-2

The rate of return of survey questionnaires was 80 percent (40 states). Not all states returning questionnaires were able to provide complete information due to limitations of their data collection. 'Total' figures for the United States were based upon projections that assumed state weights and measures regulatory program data was proportional to state populations.

State Weights and Measures Program Budgets [2002][2]

The sum of funding by sources was assumed to equal the total budget for all of the state weights and measures regulatory programs during calendar year 2002. Only 38 states, representing 86.2 percent of the U.S. population, provided complete data for this item on the questionnaire, but the combined budgets for all 50 states was based on a projection for 100 percent of the population. This projection was $135,520,462. The ratio of budget per inspector was estimated based on this information. The General Fund (44%) was equal to the next top three funding sources combined: Registration or License fees (12.7%), Inspection fees (17.4%), and Fuel Quality fees (14.2%). Clearly, the migration from the General Fund to self-supporting regulatory fees was only about 56 percent complete by 2002. No follow-on survey was conducted until 2019.

Table 1.1 State W&M Program Budgets [2002]

Funding Source	*Amount (%)*
Registration/License Fees	$14,891,449 (12.7)
Inspection Fees	20,363,985 (17.4)
Fuel Quality	16,540,049 (14.2)
Metrology Lab	1,689,881 (1.4)
General Fund	51,908,019 (44.4)
Price Verification	3,363,895 (2.9)
Other	8,061,360 (6.9)
Total (38 states)	$116,818,638 (99.9)
Projected (50 states)	$135,520,462

Notes

The 38 state budgets summarized represent 86.2 percent of the U.S. population.

See "July 1, 2002 Population" at www.eire.census.gov/popest/data/states/tables/ST-EST 2002-01.php

Table 1.2 State W&M Inspectors [2002]

W&M Inspector Ratio	*Amount*
Budget per Inspector	$73,853
Population per Inspector	157,149
Land Area per Inspector	1,928 sq. mi.
Retail Motor-Fuel Dispensers per Inspector	1,377
Scales (all kinds) per Inspector	550
Survey Count (39 states with 86.5% of U.S. population)	1,587
Projected (50 states)	1,835

Notes

Ratios based on projected total inspectors (above) and projected total RMFD and scales (all).

Land Area (2000) from www.quickfacts.census.gov/qfd/states .

State Weights and Measures Inspectors [2002][3]

It was projected that, in 2002, there were only 1,835 inspectors nationwide. It is not clear how many of these were full-time equivalent employees (FTEs). The number of administrative and management staff were not counted. It is remarkable that there was just one inspector for 157,149 population (consumers) or approximately the population of a city the size of Fort Lauderdale, Florida. Also, there was only one inspector for a land area nearly the size of the state of Delaware. As a result, inspectors spent a considerable amount of their work hours just driving to and from inspections so maintaining an annual inspection frequency for all weighing and measuring devices was simply not possible. Since inspectors only have approximately 230 work days per year (less weekends, holidays, sick days and vacation days), and there are a combined 1,927 gas pumps and scales per inspector plus considerable travel time and administrative time, it was unlikely that the average inspector had any time left over to inspect other types of weighing and measuring devices or to conduct package inspections.

Inventory of Commercial Scales and Retail Motor-Fuel Dispensers [2002][4]

Scales and gas pumps (i.e., retail motor-fuel dispensers) have been the ‘meat and potatoes’ of all weights and measures (W&M) regulatory programs from the very beginning. It was projected that, in 2002, there was a combined total three and one-half million of these devices: approximately one million commercial scales (all kinds and capacities)

Table 1.3 Commercial Scales and RMFD [2002]

W&M Devices	*Survey (37 states)*	*Projected (50 states)*
Scales (all kinds)	831,762 (82.34% of U.S. population)	1,010,155
Retail Motor-Fuel Dispensers (RMFD)	2,066,043 (81.78% of U.S. population)	2,526,343

Notes

Projection for all 50 states assumes that the total number of weighing and measuring devices is proportional to the total population. This may not be true.

Population figures for 2002 drawn from "July 1, 2002 Population" at www.eire.census.gov/popest/data/states/tables/ST-EST2002-01.php .

and two and one-half million gas pumps. Virtually all weights and measures regulatory programs expend most of their inspector resources conducting annual inspections of these devices. These devices are used to weigh (solids) or measure (liquids) commodities sold from bulk at the point-of-sale (retail). Packaged products are weighed and measured at the point-of-pack (production) using check-weighers and the net contents are displayed on the package labels.

Weighing and Measuring Device Inspection Frequencies and Failure Rates [2002][5]

The W&M devices that nearly all regulatory programs inspect on an approximately annual frequency are: gas pumps, scales (all kinds), loading-rack meters (LRM fill tank trucks that supply gas station) and vehicle-tank meters (home heating oil delivery trucks). The annual frequency largely explains the low failure rates for scales and gas pumps. Higher capacity scales generally have greater failure rates. Small computing scales fail at a rate of less than 5 percent. Survey results indicate that gas pump inspection failure rates ranged from 0.5 percent to almost 15 percent even though most were inspected nearly annually. It was believed that this variation was primarily due to differences in inspection procedures and failure criteria even though NIST *Handbook 44* recommends national standard inspection procedures and error tolerances. The "station average" policies are different among different states because there is no standard in *Handbook 44*. Station average requirements are intended to make sure that unscrupulous gas station operators do not purposely set gas pump meters at the individual tolerance of -6 cubic inches. Approximately one-half the states (22/39) fail

Table 1.4 W&M Inspection Frequencies and Failure Rates [2002]

Device Type	*Percent Inspected*	*Inspection Frequency*	*Percent Failed*	*States Reporting*
Retail Motor-Fuel Dispensers (RMFD)	83.4%	1.2 years	6.6%	39/40
Scales (all kinds)	87.5	1.1	6.9%	40/40
Vehicle-Tank Meters	95.0	1.1	12.8	22/40
Loading Rack Meters	95.6	1.0	8.8	14/40
Liquified Petroleum Gas Meters (LPG)	72.3	1.4	23.4	39/40
Railroad Track Scales	81.4	1.2	17.1	9/40
Belt Conveyor Scales	103.8	1.0	27.8	4/40
Mass Flow Meters	100.0	1.0	17.2	5/40
Taximeters	66.0	1.5	8.3	13/40
Length-Measuring Devices	92.6	1.1	8.2	7/40
Timing Devices	72.8	1.4	19.7	7/40

Notes

Formulas: % inspected = #inspected/#devices X 100, inspection frequency = # devices/# inspected, % failed = # failed/# inspected X 100.

Only RMFD, Scales, LPG and VTM data are considered reliable because more than 50 percent of states reported conducting these inspections.

a gas station for "station average" but criteria vary from -1 cubic inch error to -4 cubic inch per 5-gallon test measure or, alternately, fail a gas station if errors are predominantly minus (short measure).

The survey suggested that there was probably an optimum point for the tradeoff between inspection frequencies and failure rates. Perhaps more important is the failure history of individual gas stations. The principal lessons from the survey were that, (1) states count RMFD differently: meters (65.8%), hoses (18.4%), grade of product (10.5%) and other (5.3%); and (2) inspection procedures vary: only two-thirds of states test both slow fill and fast fill rates, less than one-half check electronic audit trails, and 5 percent of states don't test mid-grades on blenders (retail motor-fuel dispensers that mix regular and high octane – usually 50–50). These glaring differences point out the need for more uniformity in inspection procedures, failure criteria, inspector training and recordkeeping.

Device Registration and Inspection Fees [2002][6]

Regulatory fees were primarily weighing and measuring device registration or license fees (52.5%) and inspection fees (22.5%). Registration fees were generally flat fees which varied depending on the type and capacity of the device, e.g. $5 per small retail computing scale up to 30 pound capacity, $5 per retail motor-fuel dispenser up to 20 gallons per minute. Inspection fees were levied on an hourly rate. Fee revenue needed to support the state weights and measures regulatory program operating budgets determine inspection frequencies in many states rather than an optimum inspection cycle based on inspection failure histories or consumer complaint histories. Definitions of devices by capacity (e.g., small, medium, large scales) were also tied to state registration fee revenue needs. So, efforts to make device counts uniform have been frustrated by the need to change existing state laws in order to standardize definitions. And, of course, every state believes that their device definitions are the best. One-half of the states use a fiscal year (July 1–June 30) and 45 percent use a calendar year, which complicates data collection on a national basis.

Package Net Contents Inspections [2002][7]

About 90 percent (36/40) of states completing the survey reported that they conducted *some* package net contents inspections. The number of '*lots*' inspected varied considerably among these states. Firstly, the states counted inspections differently (e.g., by packages or lots or locations). The failure rate for package net contents inspections reported by *package* was 5.9 percent while the failure rate for inspections by *lot* was 12.5 percent and the failure rate by location was 20.0 percent.

In states reporting inspections by package, the total number of packages checked by these inspections ranged from 168–3,416,298 in

Table 1.5 Package Net Contents Inspections [2002]

New Contents Inspections	*States*	*Inspected*	*Failed*
by Package	18	8,859,170	522,683 (5.9%)
by Lot	15	145,755	18,203 (12.5%)
by Location	1	20	4 (20%)
Net Contents Inspections	Some Activity (<100,000)	Active (>100,000)	
by Package	11 states	8 states	
by Lot	6 states	10 states	

Notes

Only complete data reported by states was used.

The terms 'Some Activity' and 'Active' are arbitrary because state populations and the size of W&M programs were ignored.

different states and only about one-half of the states were *active*. In the states that reported by lot, the number of lots checked ranged from 484–78,422 and nearly two-thirds of states were active. The limited activity in some states indicates that most inspections are conducted on a *complaint-basis* only. The infrequent package inspections would not compel sellers to exercise more care in verifying package net contents prior to sending product to the marketplace.

Price Scanning System Inspections [2002][8]

There were problems in counting price scanning system inspections, just as with W&M device inspections and package net contents inspections. For example, the sample sizes for price scanning system inspections at small stores varied: 25 items (33.3%), 50 items (43.3%) and other (23.3%). At large stores, the sample sizes were: 50 items (30.0%), 100 items (50.0%) and other (20.0%). Another difficulty was inspection failure criteria. States failed price scanning system inspections because of either exceeding 2 percent overcharges in the sample (30.0%), 2 percent total errors in the sample (46.7%) or other criteria (23.3%). As a result, a store that passed inspection in one state may fail in another. Most packaged products are sold by regional or national chain stores so uniformity in price scanning inspections is necessary to avoid burdening industry with different, and sometimes contradictory, inspection procedures.

Conclusions from First Survey of State Weights and Measures Programs [2002][9]

The purpose of the 2002 survey was to establish baseline information about state weights and measures inspection statistics. However, it quickly became apparent that a more immediate problem was the lack of uniformity in the definition of weighing and measuring devices and inspection procedures despite the fact that the latter are included in the NIST Handbook series. According to the conclusions of the *Survey of Inspection Statistics Collected by State Weights and Measures Programs [2002]: Final Report*, "The first NCWM survey of inspection statistics collected by states helped to document the pressing need for standardized data collection practices in order to promote the goal of 'uniformity' among state weights and measures regulatory programs."[10] The stated goal of the National Conference on Weights and Measures is to promote uniformity:

> The purpose of these Uniform Laws and Regulations is to achieve, to the maximum extent possible, uniformity in weights and

> measures laws and regulations among the various states and local jurisdictions in order to facilitate trade between the states, permit fair competition among businesses, and provide uniform and sufficient protection to all consumers in commercial weights and measures practices.[11]

The survey questionnaire demonstrated that attempting to sum inspection statistics is impeded by variations among the states regarding definitions of weighing and measuring devices and inspection procedures, including failure criteria. For instance, retail motor-fuel dispensers (gas pumps) are counted variously by hoses, meters, fuel grades and dispensers. Inspection procedures for RMFD differ regarding whether to conduct both slow and fast fill tests, whether to conduct inspections of electronic audit trails of calibration mechanisms, and whether to apply wire-and-lead security seals. The 'station average' failure criteria may be applied in some states but the standard varies, e.g., -2 cubic inches, predominantly negative errors. As a result, summing inspection data for municipal, county and state weights and measures regulatory programs is tantamount to *adding apples and oranges*. Budgeting, staffing, equipping, scheduling inspections and reporting measures of effectiveness (MOE) are problematic without reliable counts of weights and measures inspection workloads and inspection results. National standards are impossible to achieve and the promotion of uniformity, by the National Conference on Weights and Measures, is defeated.

The 2002 survey report points to the need to take steps to promote uniform data collection. For example, NIST Handbook 44 *Specifications, Tolerances and Other Technical Requirements for Weighing and Measuring Devices*, must adopt standardized definitions for all weighing and measuring devices, such as, by *type* and *capacity* for scales and by *meter* for retail motor-fuel dispensers. Presently, Handbook 44 only establishes classifications of devices as needed to specify test procedures or tolerances. But, if the National Conference on Weights and Measures does not adopt standardized definitions for devices, uniformity among the state weights and measures regulatory programs simply cannot be achieved. Similarly, data collection practices must be specified in the NIST Handbooks – e.g., count '*lots*' instead of '*packages*' when conducting package net contents inspections; don't combine counts of '*inspections*' (i.e., 'as-found' condition) and '*re-inspections.*' Additionally, failure criteria (e.g., '*station average*') must be clarified and standardized inspection forms must be published in the NIST Handbooks to facilitate data collection and comparison. The 2002 survey report encouraged a shift of emphasis from '*device inspections*' to '*package inspections*' since most goods are sold in packaged form today.

Follow-Up Survey of State Weights and Measures Programs [2019]

In 2020, the author conducted a follow-up survey especially for this book. The intent was twofold: (1) to collect current data, and (2) to examine trends over time. The survey solicited weights and measures inspection data for calendar year 2019 because it was the most recent complete year of data available. Some states use a fiscal year. In that case, the fiscal year data for the fiscal years bracketing calendar year 2019 were averaged whenever possible. The COVID months of 2020 (March–December) were avoided in that process since there was an economic recession and a slow-down in government operations in most states. Unlike the 2002 state survey, the 2019 survey was not sanctioned by the National Conference on Weights and Measures (NCWM). Additionally, the 2019 survey questionnaire requested considerably less detail in order to encourage participation.

Just as in 2002, estimates of total weighing and measuring devices, device and package inspections, and inspection failures were projected from survey questionnaire responses assuming that total devices were proportional to total population. This assumption was based on the fact that the population consumes commodities sold by scales and meters. Definitions of commercial scales by category vary considerably among the states, so retail motor-fuel dispensers (RMFD) were selected as a *proxy* to represent *all* weighing and measuring devices in the nation. However, even RMFD counts submitted by states were not all in the same units. States used several methods to count RMFD including: the number of dispensers, the number of fuel grades, the number of hoses and the number of meters. Moreover, meter counts varied with some states counting a mid-grade meter for blending dispensers (e.g., a mix of regular and premium grade fuel). Ironically, many of these methods yield similar results depending on the configuration of dispenser. For example, many dispensers have a separate hose for each grade of gasoline and each grade has a separate meter except when the dispensers are blenders. Despite these caveats, the RMFD counts were still considered to be the best proxy for projecting the change in weighing and measuring devices.

The growth in the population of the United States from 2002 to 2019 was 13.5 percent (from 288,600,000 to 327,533,795). During that same period, the number of RMFD increased 16.9 percent (from 2,526,343 to 2,953,843). However, the starting figure, 2,526,343 RMFD, was itself a projection based on the 2002 state and national population figures from Census estimates. So, a direct comparison of the rate of population growth and the rate of RMFD increase was not strictly possible in order to verify the assumption that devices are proportional to population. However, projections were necessary since only 66 percent of the 50 states responded to the 2019 survey. In addition, some of that data was incomplete and, therefore, not useful. For example, several states do not maintain an inventory of weighing and measuring devices because they aren't funded by assessing a device registration fee or license fee. The budgets for those states are wholly derived from the General Fund or inspection fees that are most often charged per hour and not per device. Several other states don't record devices that fail inspection. Moreover, states that have partially privatized weights and measures (e.g., authorized service companies to perform certain device inspections) only conduct random audits so those states don't have *either* a reliable weighing and measuring device inventory or comprehensive device failure data.

State Weights and Measures Inspectors [2019]

Inspector counts for 2019 are projections that assume the number of W&M inspectors is proportional to population. The population consumes goods sold across scales or through meters. This consumption, and therefore the number of scales and retail motor-fuel dispensers, is assumed to be approximately the same per capita from state to state. States require enough inspectors to conduct inspections of nearly all scales and retail motor-fuel dispensers (RMFD) on an approximately annual cycle. The ratio of population per inspector is provided here. Ratios of commercial scales per inspector and RMFD per inspector are also shown. State inspectors are usually assigned all weighing and measuring (W&M) devices located in a specific geographical area (e.g., multiple counties) so the ratio of land area per inspector is included as well.

The substantial decreases in all of the inspector ratios shown in Table 1.6 are largely attributable to the decline in the total number of inspectors. This is concerning because inspector resources are shrinking even as inspector workloads are growing. Clearly, this is unsustainable in the long run. Hiring additional inspectors will require that W&M device registration fees be increased. The alternatives are: (1) less frequent inspections resulting in higher failure rates and more overcharges to consumers, or (2) movement towards privatization of W&M device inspections. *Privatization* will be discussed in detail in Chapter 5. However, privatization always involves an inherent *conflict of interest* because private servicing companies are for-profit enterprises that seek

Table 1.6 Total State Weights and Measures Inspectors in the U.S.

Inspector Ratios	*2002 State Survey*[a]	*2019 State Survey*[b]
W&M Inspectors	1,835	1,359 (-25.9%)
Budget/Inspector	$73,853	$120,480 (63.1%)
Population/Inspector	157,149	241,011 (+53.4%)
Land Area/Inspector	1,928 mi^2	2,774 mi^2 (+43.9%)
RMFD/Inspector	1,377	2,174 (+57.9%)
Scales/Inspector	550	749 (+36.2%)

Notes

[a] Projected from 39 states/86.5 percent of U.S. population.

[b] RMFD projected from 28 states/72.4 percent of U.S. population (50 states); inspectors projected from 26 states/68.6 percent of U.S. population; scales projected from 24 states/ 69.5 percent and budget projected from 17 states/34.2 percent of U.S. population.

Projections assume proportionality to population.

In a few instances, *extreme* outlier data provided by a state was not included in a projection to avoid skewing it.

out business and will naturally be disinclined to fail inspections and potentially lose clients.

Inventory of Commercial Scales and Retail Motor-Fuel Dispensers [2019]

There are wildly varying definitions of commercial scale categories in different states. As a result, I elected to count all scales together (all kinds) rather than attempting to group them by capacity or type. Conversely, instead of summing all meter devices together, I selected RMFD as representative of all categories of oil meters. This simplification obviously introduces errors but RMFD are the largest single category of weighing and measuring device and they are commonly found in all states.

The number of weights and measures (W&M) inspectors in the U. S. declined 25.9 percent despite a 12.3 percent *combined* growth of scales and retail motor-fuel dispensers (RMFD) from 3,536,498 (2002) to 3,971,060 (2019). See Table 1.7.

Weighing and Measuring Device Inspection Frequencies and Failure Rates [2019]

Normally, W&M device inspection frequency and W&M device failure rate are inversely related. More frequent inspections usually result in a

Table 1.7 Total Weighing and Measuring Devices in the U.S.

W&M Devices	*2002 State Survey*	*2019 State Survey*
Retail Motor-Fuel Dispensers	2,526,343[a]	2,953,843 (+16.9%)[c]
Scales (All Kinds)	1,010,155[b]	1,017,217 (+0.7%)[d]
Total W&M Devices	3,536,498	3,971,060 (+12.3%)

Notes

[a] RMFD projected from 37 states/81.8 percent of U.S. population.
[b] Scales projected from 37 states/82.3 percent of U.S. population.
[c] RMFD projected from 28 states/72.4 percent of U.S. population.
[d] Scales projected from 24 states/69.5 percent of U.S. population.

Projections assume proportionality to population.

Table 1.8 W&M Device Inspection Frequency and Failure Rate

W&M Inspections	*2002 State Survey*[a,b]	*2019 State Survey*[c,d]
Retail Motor-Fuel Dispensers	2,526,343	2,953,843
Percent Inspected	83.4%	87.4%
Frequency of Inspection	1.2 years	1.1 years
Percent Failed	6.6%	6.9%
Scales (All Kinds)	1,010,155	1,017,217
Percent Inspected	87.5%	N.A.
Frequency of Inspection	1.1 years	N.A.
Percent Failed	6.9%	N.A.

Notes

[a] RMFD projected from 39 states.
[b] Scales projected from 40 states.
[c] RMFD projected from 28 states/72.4 percent of U.S. population.
[d] No summary scale data available due to incomplete information.

Projections assume proportionality to population.

reduction in the failure rate. This would seem to be obvious on the face of it. However, RMFD inspection frequency increased slightly from 1.2 years (2002) to 1.1 years (2019) during the same period and RMFD failure rates increased from 6.6 percent to 6.9 percent. See Table 1.8. One explanation for the reduction in inspectors (discussed previously) but an increased inspection frequency is that many states have resorted to *audits* where just a portion (sample) of RMFD are inspected at locations that have maintained good compliance records. There are other explanations as well. RMFD inspection procedures have been streamlined. For example, some states only conduct a *fast fill* test and have discontinued the *slow fill* test in order to shorten times spent on

RMFD inspections. Retail motor-fuel dispensers are most likely to fail fast fill inspections and not fail slow fill inspections unless they also fail fast fill inspections. Many states have even partially privatized periodic inspections by turning them over to private service companies licensed by the weights and measures regulators. This is a growing trend. States have increasingly allowed private service companies to return W&M devices to service following passed re-inspections

Graph of RMFD Inspection Frequency v. RMFD Inspection Failure Rate [2019]

Figure 1.1 is a graph of RMFD inspection frequency v. RMFD inspection failure rate for 2019. Data from 19 states is plotted. The average (mean) inspection frequency was 1.1 years and the average (mean) inspection failure rate was 6.9 percent. Most of the states (represented as dots) are clustered about the average (1.1 yrs., 6.9%). The median is nearly the same (1.1 yrs., 6.3%) as the mean. Traditionally, weights and measures regulatory programs have sought to inspect *every device every year.* Small commercial scales and retail motor-fuel dispensers, by themselves, comprise more than 90 percent of all W&M devices and broadly impact consumption of goods by the public.

There are a few outliers but it is possible to demonstrate the tradeoff between inspection frequency and inspection failure rates. The dotted

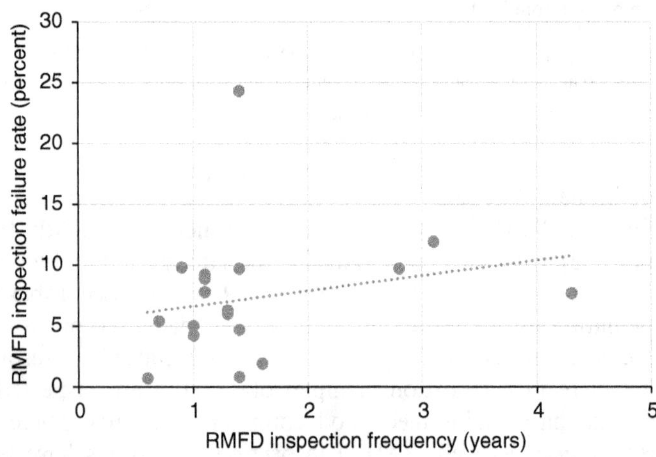

Figure 1.1 RMFD Inspection Frequency v. RMFD Inspection Failure Rate (2019).

line indicates the tradeoff. For example, if the inspection frequency was doubled to 2.2 years then the projected inspection failure rate would climb to about 8 percent and at 3.5 years the failure rate would rise to 10 percent. Historically, the overcharge errors against the public are normally about one-half of the total errors. The total overcharge to the consumers may be estimated by multiplying the average individual dispenser error recorded on the inspection report by the quantity of fuel pumped through the dispenser (totalizer reading from the dispenser) and multiplied again by the price per gallon. Performing these calculations for a small sample would allow regulators to compare the cost to the consumer of extending the inspection frequency v. the expense of hiring additional inspectors. Some argue that, since overcharges and undercharges are roughly equal, that they cancel out. The problem with that thinking is that the persons overcharged and undercharged are different and, as a result, there is no guarantee that the errors would cancel, even in the long run, if you always used the same service station. An *optimum inspection frequency* may be a longer period *if* the failure rate does not significantly increase.

Package Net Contents Inspections and Price Scanning System Inspections [2019]

Many state weights and measures programs conduct few package inspections because of lack of inspector resources. Most states have not adopted a registration fee or inspection fee revenue to fund these activities. Those states with W&M device registration and inspection fees often don't receive General Fund monies that could be used to support package inspection activities. Also, General Fund monies are too little to support regular inspections. As a result, inspectors are often limited to conducting package net contents inspections and price scanning system inspections in response to consumer complaints only.

The number of states conducting package inspection activities, such as price scanning and net contents, declined from 2002 to 2019. See Table 1.9. Price scanning inspection activities decreased from 62.5 percent (2002) of states to 54.5 percent (2019) and net contents inspection

Table 1.9 State Package Inspection Activities

Package Inspection Activity	*State Survey [2002]*	*State Survey [2019]*
Price Scanning	62.5% (25/40)	54.5% (18/33)
Net Contents	90.0% (36/40)	63.6% (2/33)

activities decreased from 90.0 percent (2002) to 63.6 percent (2019). This trend is concerning since goods of all kinds are increasingly sold in packaged form rather than from bulk using scales and meters.

One risk of state W&M programs not engaging in package inspection activities is that they will become irrelevant otherwise. Retail store owners and managers won't understand why inspectors want to conduct these inspections now because they never have before. The number of stores employing price scanning systems has spread from box stores to convenience stores and are now relied upon by virtually all retail stores. Many stores traditionally had no contact with weights and measures inspectors before they added price scanning (e.g., clothing stores, greeting card stores, dollar stores). About 80 percent of convenience stores (c-stores) sell gasoline so the *c-store* industry is already familiar with RMFD inspections.

Funding Sources for State W&M Programs [2019]

The breakdown of funding sources in the 2002 survey was by percent of state W&M budgets attributable to each source. Registration or license fees for W&M devices comprised 12.7 percent of state W&M budgets; inspection or test fees for W&M devices were 17.4 percent; and the General Fund amounted to 44.4 percent. Together, these three funding sources were 74.5 percent of state W&M budgets. Other funding sources included fuel quality (14.2 percent), metrology lab (1.4 percent), price verification (2.9 percent) and others (6.9 percent).

The state survey questionnaire in 2019 was less detailed. States were asked to identify their *predominant* funding source(s). Minor funding sources were largely ignored and the predominant funding source was used in lieu of multiple funding sources. In 2019, 54.8 percent (17/31) of states providing useable data on W&M program funding identified registration or licensing fees as their predominant funding source. Inspection or test fees were identified as the predominant funding source by 25.8 percent (8/31). The General Fund was identified by 19.4 percent (6/31) of states as the predominant funding source.

Table 1.10 State Budget Sources [2019]

Funding Source	*State Survey [2002] Percent of Budget*	*State Survey [2019] Percent of States*
Registration/License Fees	12.7%	54.8% (17/31)
Inspection/Test Fees	17.4%	25.8% (8/31)
General Fund	44.4%	19.4% (6/31)

It is evident from the changes between 2002 and 2019, that states have become largely *self-funding* (fees v. General Fund) through regulatory fee revenue. In 2019, annual W&M device registration or license fees were the predominant funding source for 54.8 percent of states and inspection or test fees were the predominant source of funding for 25.8 percent of states for a combined total of 80.6 percent. Other fees, such as service company license fees, fuel quality fees, metrology test fees and price scanning system registration fees likely comprise most of the remainder.

During the past three decades, as state budgets have become increasingly strained, any state programs that could support themselves with regulatory fee revenue were directed to do so.[12] Thus, state W&M programs have shifted funding sources from the General Fund to regulatory/license fees and inspection/test fees. Registration/license fees and inspection/test fees comprised a combined 30.1 percent of state W&M budgets in 2002 but, in 2019, these fees were the predominant source of funding for 80.6 percent of states. Likewise, 44.4 percent of state W&M budgets were from the General Fund in 2002. By 2019, the General Fund was the predominant source of funding in just 19.4 percent of states.

The ramifications of the shift from General Fund to regulatory fees to fund state W&M budgets are substantial. For example, regulatory fee revenue derived from W&M device registration/license fees and inspection/test fees must be expended only for expenses associated with those programs. As a result, package inspection regulatory programs not supported by fees, such as checking package net contents or price verification of scanning systems, are often not conducted at all. Moreover, industry lobby groups often oppose any bills at state legislatures to increase regulatory fees – seeing them as merely *taxes by another name* instead of *user fees*. Accordingly, many state W&M regulatory programs are inadequately funded causing inspector positions to go unfilled and many inspection activities to be reduced to *complaint-only* or infrequent *audits*.

Typical W&M Device Registration Fees [2019]

The range of annual W&M device registration or license fees is very broad primarily because there are no standard categories by capacity (scales) or flow rates (meters). Rather, every state has adopted its own schedule of fees. Table 1.11 illustrates ranges of *typical* W&M device registration fees. Because of the lack of standardized categories, this list is of limited use other than to compare fees in 2002 with those in 2019.

Table 1.11 Typical W&M Device Registration Fees [2019]

W&M Device	*State Survey [2002]*	*State Survey [2019]*
Scales, Small (e.g., counter)	$5	$12–35
Scales, Vehicle	$39–200	$120–250
Retail Motor-Fuel Dispensers (RMFD)	$5–12	$15–25
Vehicle-Tank Meters (VTM)	$14–75	$50–75
Loading Rack Meters (LRM)	$32–140	$75–150
Liquified Petroleum Gas (LPG) Meters	$10–150	$24–100
Taximeters	$5	$20–35

Generally, registration fees have increased across-the-board during this period. The W&M device registration fees for RMFD and small scales (e.g., counter scales used in retail) – comprising in excess of 90 percent of all commercial W&M devices – have doubled or tripled during those two decades. The increase in the *cost of living* during this period was 42.1 percent.[13] The remainder is likely attributable to the shift from the General Fund to regulatory fees as the predominant source of funding for W&M program budgets for 80.6 percent of states.

Approximately one-half of state W&M programs use device registration fees as their predominant source of funding. But about one-quarter use inspection fees (either fixed or per hour). Some states have multiple funding sources. Funds from the General Fund are fungible and can be used to subsidize the expense of conducting device inspections and allow both registration fees and inspection fees to be lower than otherwise.

Innovations to Improve Inspection Efficiency

The emphasis on W&M device regulatory fees as the predominant funding source has arguably caused state W&M programs to overemphasize device inspections at the expense of package inspection activities. For instance, many states have an *every-device-every-year* inspection philosophy because their budgets are largely funded by annual W&M device registration fee revenue. The loss of W&M inspectors has aggravated this imbalance over time. Desperate to retain an approximately annual inspection frequency for W&M devices in the face of declining inspector resources, many states have adopted innovative strategies. For example, improvements in retail motor-fuel dispenser (RMFD) inspection equipment illustrate efforts to increase the efficiency of inspectors while conducting inspections. *Prover trucks*[14] with internal storage tanks have nearly halved the time required to conduct a periodic inspection of RMFD compared with the former method of manually filling 5-gallon provers (test measures) at the pumps, moving them by welding cart then emptying them by hand into large funnels at the fill pipe of underground storage tanks. Abbreviating inspection procedures by conducting *fast fill* but not *slow fill* tests nearly halved inspection times again. *Auditing* RMFD at service stations by sampling random dispensers at one island (six dispensers) rather than all dispensers (typically 24) almost halves the inspection time again. Travel time is necessarily a part of each inspection and audits don't affect it.

Ironically, it is not uncommon for owners of commercial scales or RMFD to contact state W&M programs when annual inspections are overdue. Owners believe that they pay device registration fees for an annual check of the proper calibration of their devices and are reluctant to pay a private servicing company to check calibration on top of that. On the contrary, many device owners are unhappy with unannounced W&M device inspections by state inspectors because they find these inspections inconvenient, intrusive and expensive (to correct device calibration errors).

Annual W&M device inspections have never proved to be the optimum inspection frequency. There is a well-known inverse relationship between inspection frequency and inspection failure rate. As inspection frequency is decreased (e.g., from 1 year to 1.5 years), inspection failure rates increase. This is indicated by the positive slope of the plot of inspection frequency against inspection failure rate (see Figure 1.1). The steeper the slope, the greater the tradeoff. When the slope is flatter, the tradeoff is more favorable. In other words, the period between inspections may be extended with only small increases in the inspection failure rate. A *cost-benefit analysis* can compare the cost to the consumer of larger device errors per transaction and the benefits of redirecting inspector resources where they can have more impact.

Conclusions from Second Survey of State Weights and Measures Programs [2019]

It is very clear from both state surveys that all states should add uniform weights and measures device categories to existing databases so that data from different states may be summed. In other words, convert the data so that all comparisons are '*apples and apples*.' No state laws need to be changed to do this – the usual excuse offered for not adopting uniform device categories. This would be a simple project to set up and maintain. It could be automated. Only basic data is needed: (1) inventory of weighing and measuring devices by category (e.g., from registration fee records or inspection records); (2) record of inspections/audits conducted; and (3) record of inspection failures including device errors (as found). Inspection forms and computer templates should be modified to record the standard device category data. Many states keep incomplete records that are not very helpful for planning or analysis of inspection activity. Without a *feedback loop*, there is no method to determine whether the effectiveness or efficiency of the weights and measures regulatory program may be improved. It is important that state weights and measures programs issue annual *report cards* on all inspection activities and that historical data is compared over time to assure that there is constant improvement. The National Conference on Weights and Measures (NCWM), in conjunction with the National Institute of Standards and Technology/Office of Weights and Measures (NIST/OWM), should collect state data annually, using standard categories, to assess whether changes are needed in uniform inspection procedures contained in the NIST Handbook (series) or inspector training.

At the 92nd National Conference on Weights and Measures at Salt Lake City, Utah in 2007, the NCWM adopted standard categories of weighing and measuring devices to facilitate data collection and analysis from the various state weights and measures programs.[15] The device codes adopted by the NCWM are summarized in Table 4-12.

> The NCWM has recommended a categorization scheme for measuring instruments that jurisdictions are encouraged to use, either by revising their current device categories or supplementing their current categories with an additional category structure in their database that will allow comparison on a national basis. The device registration programs of many states are based on different categories of devices, so it is difficult to change the device registration categories. However, by adding another category structure to their database, states will ultimately be able to compare inspection results across the country.[16]

Unfortunately, the use of the 'NCWM Device Category Codes' was made voluntary and, judging from the data submissions for the 2019 state survey, the initiative has not been implemented. Until all states use uniform weighing and measuring device category codes, no real progress can be made to collect accurate data nationwide. It is ironic

Table 1.12 NCWM Device Category Codes

Device Code	*Category*	*Capacity*	*Examples*
SP	Scale, Precision	< 5 g. scale division	jewelry, prescription scales
SS	Scale, Small	< 300 lbs.	retail computing scales
SM	Scale, Medium	301 to 5,000 lbs.	dormant, platform scales
SL	Scale, Large	>5,001 lbs.	livestock, recycler, hopper, belt conveyor
SV	Scale, Vehicle	>40,000 lbs.	vehicle, railway track scales
MS	Meter, Small	<30 gpm [a]	retail motor-fuel dispensers
MM	Meter, Medium	30 to 200 gpm	vehicle-tank meters
ML	Meter, Large	>200 gpm	loading rack, agri-chemical, bulk oil meters
MF	Meter, Mass Flow	All	heated tanks of corn syrup (soft drinks)
MW	Meter, Water	All	water sub-meters for apartments, mobile homes
MG	Meter, LPG	All	propane
MT	Meter, Taxi	All	taximeters
DT	Device, Timing	All	clocks in parking lots
DL	Device, Length	All	cordage meters
GM	Grain Moisture	All	
GA	Grain Analyzer	All	
MD	Multiple Dimension Measuring Device	All	
MC	Meter, Cryogenic	All	

Source: NIST Special Publication 1070 *Report of the 92nd National Conference on Weights and Measures: Salt Lake City, Utah – July 8 through 12, 2007 (November 2007)*, Item 402–2 "Standard Categories of Weighing and Measuring Devices, p. PDC-8.

that the principal goal of the NCWM and NIST/OWM – to promote uniform standards for weights and measures – is defeated before it can begin simply for the lack of standardized device categories.

State Weights and Measures Regulatory Organizations

State weights and measures regulatory programs are fairly similar in terms of inspection activities. The principal differences lie in how they are organized (i.e., who conducts the inspections): (1) *state only* (e.g., Oregon, Alaska – state inspectors only), (2) *state-local hybrid* (e.g., Washington – state and Seattle inspectors, California – state and county inspectors). The weights and measures regulatory programs in large states tend to be '*state-local hybrid*,' for example, New York, Pennsylvania and Ohio. Hybrid states also include smaller states such as Massachusetts.

Some states have been experimenting with 'privatization' of weighing and measuring device inspection activities. New Hampshire has privatized inspection of all weighing and measuring devices but conducts audits and investigates complaints. Kansas has privatized inspection of commercial scales only. Many states permit licensed service companies to place a device back in service after it passes re-inspection. The topic of '*privatization*' will be discussed in considerable detail in Chapter 5.

All states maintain responsibility for oversight of local government (county and municipal) weights and measures regulatory programs. Normally, state weights and measures divisions are located within state departments of agriculture. This can be problematic because state weights and measures are, essentially, the '*quantity police*' while the department of agriculture, in most states, exists largely to *promote* the agriculture industry. County and municipal legal codes adopt state code so all weights and measures enforcement activities are uniform with a state.

Some states have moved fuel meter inspection to another state agency but, typically, both weighing devices and liquid metering devices are regulated by the same agency.

Metrology Laboratories

States operate metrology labs to periodically calibrate all field standards (e.g., test weights, test measures) used by weights and measures inspectors to conduct inspections. The labs also calibrate standards submitted by various industries in the states, which require a high degree of precision in their manufacturing processes. Finally, some labs perform testing of prototype weighing and measuring devices in order to verify compliance with the applicable section of Handbook 44 as part of the National Type Evaluation Program (NTEP) supervised by the National Conference on Weights and Measures (NCWM).

Examples of some weighing and measuring equipment calibrated by metrology laboratories include: liquid test measures (e.g., 300-gallon in-ground and portable (trailer-mounted) provers, 5-gallon hand provers, calibrated glassware); test weights (e.g., 25 and 50-pound hand weights, case weights, air weights); calibrated stop watches and length-measuring devices. State laboratories are operated by metrologists. The laboratories are granted certificates of metrological *traceability* by NIST/OWM in recognition that the state standards are traceable (calibrated) to national and international standards. The purpose of calibrating *field* standards used by inspectors is to assure '*traceability*' through state reference standards and back to the national reference standards at NIST in Gaithersburg, Maryland. Laboratories are also accredited by the National Voluntary Laboratory Accreditation Program (NVLAP), administered by NIST, if their operating procedures meet established criteria. Inspectors do not only conduct inspections to verify that commercial weighing and measuring devices meet tolerances for accuracy, but also to verify compliance with all other requirements in NIST Handbook 44. The NTEP program issues a Certificate of Compliance (CC) to manufacturers of devices that pass testing by designated metrology laboratories. During field inspections, CC numbers on equipment label plates are checked by inspectors.

Notes

1 National Conference on Weights and Measures. "Survey of Inspection Statistics Collected by State Weights and Measures Programs [2002]: Final Report" NCWM-News, 2003, Issue 3, p. 10.
2 Ibid., Table 1 "State Weights and Measures Program Budgets [2002]", p. 10.
3 Ibid., Table 2 "State Weights and Measures Inspectors [2002]", p.10.
4 Ibid., Table 3 "Inventory of Commercial Weighing and Measuring Devices [2002]", p. 11.
5 Ibid., Table 4 "Weighing and Measuring Device Inspection Frequencies and Failure Rates [2002]", p.11.
6 Ibid., p. 12.
7 Ibid., p. 13.
8 Ibid., p. 12.
9 Ibid., pp.13–14.
10 Ibid. p. 13.
11 National Institute of Standards and Technology. NIST Handbook 130. *Uniform Laws and Regulations: in the areas of legal metrology and engine fuel quality.* Section I "Introduction", B. "Purpose", p. 1.
12 Regulatory fees are intended to support a specific regulatory program or activity. These fees cannot be used to subsidize other programs. Revenue fees are intended to raise revenue for the General Fund to support other programs or projects.
13 Consumer Price Index for all urban consumers (CPI-U) 1984=100. Dec 2002 (179.9), Dec 2019 (255.7).
14 Prover trucks usually have three installed 5-gallon liquid test measures (provers) that drain into integral (built-in) independent tanks designed to keep the grades separate. The tanks are gravity drained back to underground storage tanks at the conclusion of the inspection.
15 NIST Special Publication 1070 *Report of the 92nd National Conference on Weights and Measures: Salt Lake City, Utah – July 8 through 12, 2007*

(November 2007), Item 402–2 "Standard Categories of Weighing and Measuring Devices, pp. PDC-8 and PDC-9.

16 National Institute of Standards and Technology. Handbook 155. Weights and Measures Program Requirements: A Handbook for the Weights and Measures Administrator (2011), p. 80.

Bibliography

National Conference on Weights and Measures. "Survey of Inspection Statistics Collected by State Weights and Measures Programs [2002]: Final Report." NCWM-News, 2003, Issue 3.

National Institute of Standards and Technology. *NIST Handbook 130. Uniform Laws and Regulations: In the Areas of Legal Metrology and Engine Fuel Quality.*

National Institute of Standards and Technology. *Handbook 155. Weights and Measures Program Requirements: A Handbook for the Weights and Measures Administrator* (2011).

National Institute of Standards and Technology. *NIST Special Publication 1070. Report of the 92nd National Conference on Weights and Measures: Salt Lake City, Utah – July 8 through 12, 2007 (November 2007).*

2 The Economics of Weights and Measures Regulation in the United States

There has not been any extensive research on the economics of weights and measures regulation in the United States. This chapter is an attempt to present what we know or suspect on that subject. Let's start with the big picture (macroeconomics) and then move to individual markets (microeconomics).

Gross Domestic Product (GDP)

The size of the economy of the United States is most often measured as the Gross Domestic Product (GDP). The common method of computing GDP is to sum all spending on goods and services by the final users. The formula is expressed as follows: $GDP = C + I + G + NetEx$ where '*C*' is consumer spending on durable and nondurable goods and services or *Personal Consumption Expenditures (PCE)*; '*I*' is business spending or *Gross Private Domestic Investment*; '*G*' is government spending or *Government Consumption Expenditures and Gross Investment*; and '*NetEx*' is net exports or *exports less imports*. The '*current-dollar GDP*,' or *nominal GDP*, for 2019 – that is, GDP not indexed to a base year to remove inflation – is shown in Table 2.1.[1] The source is the Bureau of Economic Analysis (BEA) in the Department of Commerce.

Impact of Weights and Measures Regulation on GDP

As you can see, current dollar GDP in 2019 was nearly $22 *trillion* dollars. Consumer spending (C), by itself, is normally 65–70 percent of annual GDP. The Office of Weights and Measures in the National Institute of Standards and Technology (NIST/OWM) has estimated that weights and measures regulation impacts approximately 50 percent of GDP.

DOI: 10.4324/9781003263661-3

Table 2.1 Gross Domestic Product (GDP) – 2019

Component of GDP	*Amount (Billions of Dollars) and Percent of GDP*
C, consumer spending	$14,759.2. (67.9%)
I, business spending	3,732.6 (17.2%)
G, government spending	3,805.3 (17.5%)
NetEx, net exports	(549.8) (-2.5%)
Gross Domestic Product (GDP)	$21,747.4

> Weights and measures activities are pervasive within the United States. It is estimated that U. S. weights and measures regulations impact roughly half of the U.S. gross domestic product. The success of the commercial measurement system can be judged by the ease with which transactions are executed, the level of confidence that buyers and sellers have, and the accuracy with which these transactions are performed.[2]

Weights and measures regulation prescribes the method of sale for most goods and services. Goods are offered for sale by weight, liquid or dry measure, count and other quantity measures. Some goods (e.g., apparel) are not weighed or measured but are sold by the *each* through electronic price scanning systems. Generally, commodities in packaged form are measured at point-of-pack (production) and commodities sold from bulk are measured at point-of-sale (retail). Services are often sold by time using measuring devices such as taximeters (distance, time), parking meters and laundromat appliance timers. The NIST/OWM aptly described the diversity of goods and services sold in the commercial measurement system in NIST *Handbook 155*, a guide for weights and measures administrators.

> Many commercial transactions are based on weight, volume, length, or count of products bought and sold. Packaged goods are purchased at the supermarket, people buy delicatessen items over price computing scales, gasoline and diesel fuel are purchased through pumps (retail motor fuel dispensers), gasoline and diesel fuel must meet prescribed quality or octane standards, scanners are used at checkout stands in retail stores to look up prices of products identified by bar codes, farmers sell grain, produce and livestock over scales, grain prices are adjusted up or down based upon quality measurements [e.g. moisture meters], and landfills charge fees based upon the weight of trash delivered. The structure within

> which transactions among businesses and with the general public are conducted is called the commercial measurement system.[3]

No report on the percent of GDP affected by weights and measures regulation has been issued by NIST/OWM, but Henry Opperman, then Chief of the Weights and Measures Division at NIST, provided a brief summary in 2005.

> People and businesses buy and sell products every day, with most products weighed, measured, or sold in packages that declare the net content. 'Weights and measures' is the term applied to the rules, regulations, and standards that govern the measurements of the quantities for those transactions to ensure equity in the marketplace. Based on the most recent available (1998) data, it is estimated that weights and measures regulations affect about $4.5 trillion of transactions, representing approximately 52 percent of the U.S. gross domestic product. Food and beverage stores, which are a major focus of weights and measures oversight, had over $487 billion in sales in 2002 according to U.S. Census Bureau Economic Survey statistics.[4]

If the NIST/OWM estimate regarding the impact of weights and measures regulation on GDP is correct at approximately 50 percent, then the impact in 2019 would have been $11 trillion compared with $4.5 trillion in 1998. This is nearly a tripling of the value of goods and services affected by weights and measures regulation in just 21 years. During 2002–2019, roughly the same period, the number of weights and measures inspectors declined 25.9 percent. As a result, the capacity of state and local weights and measures programs to effectively enforce weights and measures regulations has become increasingly problematic.

An estimate of the benefits from weights and measures inspections in the United States, based on similar work in Canada, was offered by Dr. William A. Jeffrey, NIST Director, in his President's Address to the National Conference on Weights and Measures at Chicago, Illinois on July 11, 2006:

> Analyses done in the United States and in other developed economies demonstrate that weights and measures underpin transactions that account for over half of the GDP.
>
> And continued verification of the accuracy is critical. In 1997 a Canadian case study reported that, on average, each weights

and measures inspector discovered and corrected about $2 million worth of 'measurement inequity.'

Unfortunately, we don't have a comparable estimate for the United States. But we do know that in 2002, the median investment in state Weights and Measures operations was about $50,000 a year for each inspector.

If we assume similar levels of performance in the United States as in Canada – about $2 million in benefits per inspector – then a rough estimate of society's rate of return would be 40 to 1 – a phenomenal result.[5]

The Expenses of Weights and Measures Regulation

A primary concern of regulated businesses is the expense of complying with regulatory requirements. The negative slope of the demand curve indicates that these expenses are shared by consumers and businesses alike. In order for businesses to pass through *all* regulatory expenses, the demand curve would have to be vertical or '*perfectly inelastic*.' That means consumers would demand the same quantity no matter what the price.

The demand curve plots price (y axis) against quantity demanded by consumers (x axis). The quantity demanded declines if prices are increased (negatively-sloped demand curve). In other words, the quantity demanded by consumers is less when products cost more. Conversely, the quantity supplied increases if prices are increased so the supply curve is positively sloped. There is an equilibrium price and quantity at the intersection of both curves. There is no surplus of unsold product (e.g., more quantity supplied than the quantity demanded at a given price) at the equilibrium price, but there may be a shortage in the market at prices lower than the equilibrium price.

A change in product price results in different quantity demanded and is represented by *movement along* the curve. Changes in non-price determinants of demand or supply cause a *shift* of the curves. For example, the additional expense of compliance with new regulatory requirements causes an upward shift of the supply curve, from S_0 to S_1, leading to a higher equilibrium price (P_1) at the intersection of D_0 and S_1. As a result, the quantity demanded is reduced from Q_0 to Q_1. The consumer absorbs a portion of the price increase because the demand curve (D_0) is not horizontal. Still, regulatory expenses reduce the profitability of businesses.

The graph, Figure 2.1, considers a *perfectly competitive market structure*, or a marketplace with a large number of small companies so that

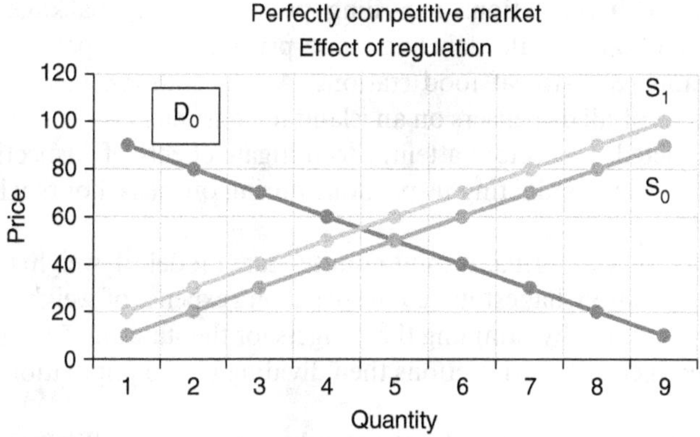

Figure 2.1 Effect of Regulatory Compliance Expenses on Product Price and Quantity Demanded.

no one company can influence the market price. The price-quantity model is different with different market structures: perfect competition, monopolistic competition (i.e., perceived product differentiation due to advertising), oligopoly (few sellers) and monopoly (one predominant company). Most retail markets offering goods or services to consumers are perfect competition or monopolistic competition.

Some weights and measures regulatory expenses are rather minor such as licensing or inspection fees. Regulatory requirements that affect the packaging of products or the method of sale may have a more substantial impact. For example, unit price regulations require food stores to display shelf tags stating the price per pound or price per gallon so that consumers may make value comparisons between products in the same package size from different producers or different package sizes by the same producer. Supermarkets stocked an average of 28,112 items in 2019[6] so the management of unit price shelf tags requires store computers and staff hours that could amount to a significant ongoing cost. This is particularly true since the average net profit after taxes for supermarkets was just 1.0 percent in 2019.[7] Meat departments in supermarkets must affix labels on store-packed meat indicating, apart from other information, the net weight of each package. Net weight is the gross (package) weight less the weight of the tray, wrapper, soaker pads, loose, clear fluids (purge) or, in other words, all *tare*. The cost of weighing and labeling each random pack (i.e., each package has a different weight) meat product may be substantial. In addition, periodic unannounced inspections at grocery stores take up time for store staff.

Package inspection activities include, but are not limited to, checking compliance with regulations regarding price scanning, package net contents, method of sale, labeling, unit pricing and proper tare at checkout and ready-to-eat food stations. At gas stations, inspections close one side of all dispensers on an island to consumers while tests are being conducted. Inspectors attempt to mitigate effects of inspections at retail stores by conducting inspections during off-peak hours whenever possible.

The State of California, a joint county-state model of weights and measures regulation, has estimated the per capita expense of weights and measures regulation by summing the budgets for the state and 55 county weights and measures jurisdictions then dividing by the population.

> The cost per capita includes the state's and counties' efforts ($0.17 per person per year, and $1.32 per person per year, respectively) for a total of $1.49 per person per year. California consumers are protected from intentional or unintentional fraud and unfair business practices, and California businesses may operate confidently in a marketplace that is fair, transparent, competitive, and equitable for each business owner.[8]

Most regulation of businesses for the benefit of consumers takes the form of consumer protection or product safety. There are agencies in federal, state and local government that are responsible for specific types of regulation. Weights and measures regulation is principally for the purpose of consumer protection or making sure that the consumer gets what he or she pays for. This means accurate quantities. Weights and measures regulation also helps maintain a *level playing field* between competing businesses by assuring that all business *play by the same rules*. Weights and measures inspectors can be seen as the *quantity* police. The quality or safety of products is the responsibility of other regulatory agencies.

Weights and measures regulation is a law enforcement activity. Inspectors conduct unannounced inspections then take enforcement action as indicated. This includes civil penalties (fines); orders taking product off sale; orders taking weighing or measuring devices out of service or condemning them; and suspending or revoking licenses among other actions. Sometimes businesses enter into settlement agreements with regulators to avoid prosecution. For example, in the case of repeated price scanning system failures, a business may agree to take specific steps such as conducting frequent self-audits, posting signs for consumers notifying them how to make complaints to weights and measures officials and assigning a pricing manager. Inspectors may conduct training on compliance with applicable weights and measures regulations for store staff, or at periodic manager or store meetings. Voluntary compliance is *always* preferred to taking enforcement action.

Economic Impacts of Regulation

Many states do not conduct analysis of 'as found' device calibration errors from annual inspections despite the fact that this information is crucial for estimating economic losses borne by consumers as a result of short measure transactions. For example, if a grade of fuel on a fuel dispenser is tested and the sight glass on the 5-gallon test measure indicates an -8 error for regular gas (a minus error is short measure), then the shortage is 8 cubic inches on a 5-gallon draw. This represents a shortage of -8 cubic inches/1,155 cubic inches per 5-gallon test measure or 0.7 percent on all regular grade gasoline passing through that fuel dispenser. If that grade on that fuel dispenser sells an average of 500 gallons per day, the shortage would amount to approximately 3.5 gallons per day or $10.50 at $3.00 per gallon. Annually, the overcharges would sum to $3,832.50 for just one fuel grade on the fuel dispenser if no service agent visits for a year to check calibration:

> -8 cubic inches per 5 gallons X 500 gallons per day = 800 cubic inch shortage per day (volumetric measure)
> -800 cubic inches per day/ 231 cubic inches per gallon = 3.5 gallons shortage per day (liquid measure)
> 3.5 gallons per day X $3.00 per gallon = $10.50 per day overcharge
> $10.50 per day X 365 days per year = $3,832.50 annual overcharge

At first, an error of less than 1 percent (0.7%) appears relatively insignificant. However, the cumulative effect of even a small calibration error adds up quickly. If it is assumed that the gas station owner pays a device registration fee of $10 annually for one gas pump, then the 'cost: benefit' of the inspection for the consumer is $10: $3,832.50. Approximately one-half of fuel dispenser errors are undercharges to the consumer. If this case were an undercharge (+8 cubic inches per 5-gallon test measure), the gas station owner will save $3,832.50 on just this one grade on one fuel dispenser alone. All calibration errors are a cost against either the consumer or the business. This is very strong evidence for the economic efficacy of weights and measures regulation.

In this example, the cost of regulation ($10 annual device registration fee) would actually be split between the buyer and seller due to the negatively sloped demand curve. The full amount of the regulatory fee cannot be passed through to the consumer by the seller unless the demand curve is perfectly inelastic (vertical) – which does not exist in the marketplace.

One powerful argument against using the 'inspection frequency: inspection failure rate' tradeoff criteria *by itself* to justify a decision to lengthen inspection frequencies to reduce weights and measures regulation expenses is contained in the preceding example. If the inspection frequency was increased from one year to two years, and there was no annual service company visit in the interim, then the economic loss to the consumer – or the business – would be doubled to $7,665. This is a strong argument for more frequent rather than less frequent inspections.

Inspection records normally reveal that the number of overcharges and undercharges are roughly the same. Despite the fact that most overcharges are not intentional, the overcharge still harms the consumer. In the foregoing example, the -8 error is a typical error found in routine inspections. The allowed tolerance on fuel dispenser over-registration (overcharge) on quantity dispensed is minus 6 cubic inches so the failed fuel dispenser is only slightly out of tolerance. But even a small error is multiplied over many transactions *and* until it is corrected. As you can see, the old weights and measures saying that, 'Weights and measures saves millions of dollars, one nickel at a time,' accurately depicts the *multiplier effect* of device errors in the marketplace on both consumers and businesses.

Notes

1 Bureau of Economic Analysis (BEA), U.S. Department of Commerce. Table 1.1.5 "Gross Domestic Product" (Rev. February 25, 2021). https://apps.bea.gov. Accessed March 23, 2021 11:55 PM.

2 National Institute of Standards and Technology (NIST), U.S. Department of Commerce. *Weights and Measures Program Requirements: A Handbook for the Weights and Measure Administrato*r. NIST Handbook 155 (2011), Sec. 1.0 "The Commercial Measurement System." p. 1.

3 Ibid.

4 Henry Oppermann. "NIST's Role in Weights and Measures" *ASTM Standardization News*. (January 2005), p. 1/ 4. www.astm.org/SNEWS/January_2005/oppermann_jan05.html . Accessed March 26, 2021.

5 National Institute of Standards and Technology. Special Publication 1053. *Report of the 91st National Conference on Weights and Measures*. (2006), p. GS-3. "President's Address to the National Conference on Weights and Measures at Chicago, Illinois on July 11, 2006 by Dr. William A. Jeffrey, NIST Director."

The ratio of weights and measures budgets per inspector in 2002 was $73,853 according to the NCWM survey of state weights and measures programs conducted that year. See Chapter 4, Table 4.1.

6 The Food Industry Association/Food Marketing Institute (FMI), "Supermarket Facts." www.fmi.org/our-research/supermarket-facts

7 Ibid.

8 California Department of Food and Agriculture. *Annual Report to the Legislature: Division of Measurement Standards. Fiscal Year 2018/19*, p. 7.

Bibliography

California Department of Food and Agriculture. *Annual Report to the Legislature: Division of Measurement Standards. Fiscal Year 2018/19.*

Henry Oppermann. "NIST's Role in Weights and Measures" *ASTM Standardization News* (January 2005).

National Institute of Standards and Technology. Special Publication 1053. *Report of the 91st National Conference on Weights and Measures* (2006).

National Institute of Standards and Technology (NIST), U.S. Department of Commerce. *Weights and Measures Program Requirements: A Handbook for the Weights and Measure Administrator. NIST Handbook 155* (2011).

3 Case History

Weights and Measures Regulation in Seattle, Washington

In this chapter, an extended example is provided to illustrate the nature of weights and measures regulation by a *typical* municipal weights and measures jurisdiction. Although municipal weights and measures regulatory programs can be quite different than state programs, all weights and measures programs share quite a bit in common. This case history is about Seattle, Washington since my experience and knowledge is based on a 20-year career as the manager of the weights and measures regulatory program there. As a result, unlike the rest of this book, a portion of Chapter 3 is written in the first person (i.e., 'I') where the author is writing from his own experience.

The Early Years

Establishment of Weights and Measures Division in Seattle

Weights and measures in Washington State was originally a combination of large city and county programs. The first weights and measures regulatory program in Washington began in the city of Spokane. Shortly afterwards, other programs began in Seattle, Everett and Tacoma. Approximately 25 rural counties shared weights and measures officials.[1] The state established a weights and measures regulatory program in 1913, within the Department of Agriculture, after hiring Seattle's first chief inspector of weights and measures.[2] The state weights and measures program gradually expanded as cities and counties closed their local programs due to budget shortfalls. By the late 1990s, the Tacoma program closed leaving just the state program and municipal programs in Seattle, Everett and Spokane. Currently, besides the state weights and measures regulatory program, only Seattle has an active municipal program.

DOI: 10.4324/9781003263661-4

The weights and measures regulatory program in Seattle was established by Ordinance No. 26018 signed by the Mayor on December 31, 1910 (the '*1910 Ordinance*').

> ORDINANCE NO. 26018. An ordinance relating to the weighing, measuring and inspecting of all commodities sold or offered for sale by weight or measure within the City of Seattle; to enforce the keeping of proper legal weights and measures by all vendors in the City; to provide for the inspection thereof, inspection fees therefor, and the issuance of licenses therefor; and providing penalties for violation thereof.

Staffing for the new Seattle Weights and Measures Division was authorized by Ordinance No. 26477 and signed by the Mayor on February 28, 1911.

> BE IT ORDAINED BY THE CITY OF SEATTLE AS FOLLOWS:
>
> Section 1. That the Superintendent of Public Utilities hereby is authorized

Photo 3.1 Inspector with cart of 50-pound test weights for large scale (c. 1916).

Source: Courtesy of Seattle Municipal Archives [Identifier 184105].

> To employ in the Weights and Measures Division of the Public Utilities Department, subject to Civil Service Rules and Regulations, the following employees whose salaries shall be as indicated:
>
> One (1) Inspector of Weights and Measures at a salary not to exceed One Hundred and Twenty-five ($125.00) Dollars per month;
>
> One (1) Assistant Inspector of Weights and Measures at a salary not to exceed One Hundred and Fifteen ($115.00) Dollars per month;
>
> One (1) Assistant Inspector of Weights and Measures at a salary not to exceed One Hundred ($100.00) Dollars per month; and
>
> One Book-keeper and Stenographer at a salary not to exceed Eighty-five ($85.00) Dollars per month.

A.L. Valentine, Superintendent of Public Utilities, appointed Arthur W. Rinehart as the first Chief Inspector of the Weights and Measures Division on April 20, 1911.[3] Chief Inspector Rinehart hired Leslie J. Allen, previously employed as an accountant, as his First Assistant Inspector on May 15, 1911.[4] The Department of Public Utilities was comprised of three divisions: Public Utilities Division, Municipal Street Railway Division and Weights and Measures Division.

Photo 3.2 Weights and Measures Division staff (c. 1920).
Source: Courtesy of Seattle Municipal Archives [Identifier 191663].

Photo 3.3 Superintendent of Public Utilities (c. 1911).

Source: Seattle Consumer Affairs Unit. *Seattle Weights and Measures Regulation 1911–2011: A Century of Consumer Protection* (March 26, 2011).

Establishment of Weights and Measures Department in Washington

On March 11, 1913, Governor Ernest Lister signed Senate Bill 61 creating the new Washington State Department of Weights and Measures. Subsequently, Arthur W. Rinehart was hired by the State of Washington

Photo 3.4 First Chief Inspector of Seattle Weights and Measures Division (1911) and first Chief Deputy of new Washington State Department of Weights and Measures (1913).

Source: Seattle Consumer Affairs Unit. *Seattle Weights and Measures Regulation 1911–2011: A Century of Consumer Protection* (March 26, 2011).

to become the first Chief Deputy of the new state Department of Weights and Measures on May 31, 1913. Leslie J. Allen was appointed to replace him as the Chief Inspector of the Weights and Measures Division in Seattle.[5]

Weights and Measures Standards

The new Weights and Measures Division for the City of Seattle did not become active until it received weights and measures *standards* from the National Bureau of Standards.[6] These standards were used to calibrate field standards used by inspectors to test the accuracy of scales and meters used by vendors. Later, the state established a metrology lab and used its standards to calibrate all field standards used in Washington. State standards were calibrated against the national standards at the National Bureau of Standards so Seattle standards were, ultimately, '*traceable*' back to the national standards.

Organization of Seattle Weights and Measures Division

Seattle weights and measures inspectors were delegated the powers of *special policemen* by the Chief of Police and carried a badge [Seattle Ordinance No. 26018, Sec. 2]. The inspectors were empowered by the 1910 Ordinance to assess fines or make arrests for violations of weights and measures law. Fines ranging from $5 to $100 and up to 30 days confinement in the city jail were provided for convictions. [Sec. 23]. Inspectors were required to conduct annual inspections on weights and measures used in commerce. Weighing and measuring devices (scales and meters) could not be used by vendors unless the calibration mechanisms were sealed by inspectors.

Scales that could not be corrected within ten days were condemned, confiscated and destroyed so they could not be used. [Sec. 22]. Periodically, condemned weighing and measuring devices and short-measure metal milk bottles (dented) were taken out into Elliot Bay and dumped in deep water.

The 1910 Ordinance also required that certain packaged food items must display the net weight. In 1913, the State Department of Weights and Measures was successful in passing a new state law through the legislature which required that certain commodities were to be sold in standard packages – vinegar, milk, butter, bread, berries, coal in sacks and potatoes in sacks.[7]

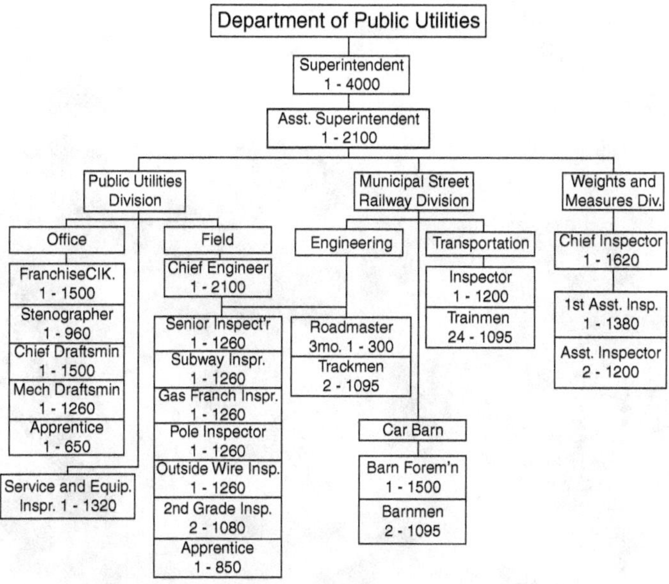

Figure 3.1 Organization chart for the Seattle Department of Public Utilities City of Seattle.

Source: Annual Report of the Department of Public Utilities (1913).

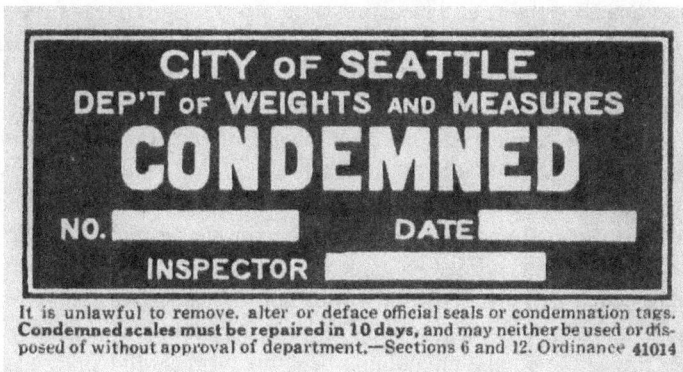

Photo 3.5 Early 'Condemned' decal used by City of Seattle.

Source: Photograph by author.

Photo 3.6 Seattle weights and measures inspectors with condemned scales (c. 1915).

Source: Courtesy of Seattle Municipal Archives [Identifier 184118].

The Seattle weights and measures regulatory program was funded by *inspection fees* assessed for commercial weighing and measuring devices. For example, a fee of $0.50 was assessed for scales with capacities of 35 pounds, which were commonly used to weigh produce in public markets. Today, Pike Place market is a well-known tourist attraction downtown at the foot of Pike Place. In the early twentieth century, there were many public markets because most fruit and vegetables consumed by the public were offered for sale in public markets by local farmers. The inspection fee for gasoline fuel dispensers was $0.25. The inspection fees were removed in 1912 and money was appropriated from the General Fund to maintain the Seattle Weights and Division. It was determined that, “the whole people are equally benefited and should share equally in the cost.”[8]

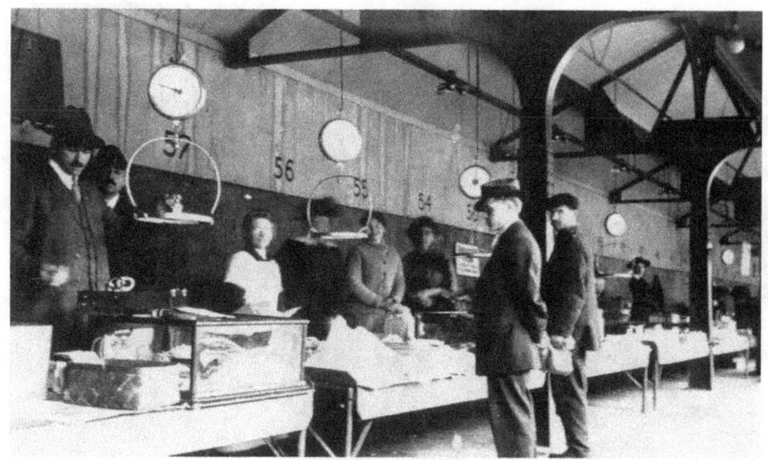

Photo 3.7 Farmers Market on Pike Place near the waterfront (c. 1916).
Source: Courtesy of Seattle Municipal Archives [Identifier 35935].

Seattle Weights and Measures Division Inspection Activities During the First Year

The first full year of weights and measures regulation was 1912. One of the principal inspection activities conducted that year was verifying horse-drawn wagon loads of coal widely used for heating and cooking. Inspectors checked 151 wagon loads. According to Chief Inspector Rinehart, "Seattle people use approximately 720,000 tons of coal per year. At an average cost of $5 per ton, a shortage of 50 pounds on each ton would mean a total loss of $90,000."[9] Inspectors tested 2,605 scales of all kinds, approving 1,928 and condemning 677.

Only 47 gasoline dispensers were tested with 34 approved and 13 condemned. There were 52 arrests resulting in 18 convictions and $255 in fines.[10] Seattle households consumed approximately 6,000,000 pounds of butter annually. Package inspections were largely bottles of milk, sacks of potatoes, loaves of bread and blocks of butter.[11]

There is a quaint story about weights and measures inspectors investigating the practice of 'cutting out' by icemen. In the early twentieth century, families purchased 25-pound blocks of ice to keep food cool in the ice box (predecessor of the electric refrigerator). The ice was delivered by icemen who drove horse-drawn ice wagons through

Photo 3.8 Seattle weights and measures inspector checking coal wagon.
Source: Seattle Consumer Affairs Unit. *Seattle Weights and Measures Regulation 1911–2011: A Century of Consumer Protection* (March 26, 2011).

Photo 3.9 Seattle weights and measures inspector filling 5-gallon test measure.
Source: Seattle Consumer Affairs Unit. *Seattle Weights and Measures Regulation 1911–2011: A Century of Consumer Protection* (March 26, 2011).

residential neighborhoods. Icemen purchased 200-pound cakes of ice from ice companies for $5 per ton then sold it to households in 25-pound blocks. They used ice picks to cut the ice cake and carried ice scales to weigh the ice blocks. Sometimes icemen cut the 200-pound cake into nine blocks to earn a little more. Often children chased the ice wagons through the neighborhoods asking for ice chips to suck on during the hot summer days.[12]

Early Seattle Weights and Measures Division Inspection Activities

Weights and measures inspections during the first four years, 1912–1915, are summarized in the table below.

The emphasis of firewood inspections was to assure that cut wood was sold by the '*cord*' or 128 cubic feet (4 feet x 4 feet x 8 feet). Package inspections were focused on checking that the net contents declared on a label was accurate and that certain commodities were offered for sale in standard packs (e.g., gallon of milk, pound of butter). This practice, for the most commonly consumed commodities, allowed the public to make value comparisons to determine the least expensive.

Not included in the preceding inspection activity summary, until 1915, are miscellaneous activities such as taxi-meter tests [Ordinance

Table 3.1 Weights and Measures Inspection Activity 1912–1915 in Seattle, Washington

Inspection Activity	*1912*	*1913*	*1914*	*1915*
Scales tested	2,605	3,333	3,879	3,136
[Approved/ Condemned]	[1,928/677]	[2,749/584]	[2,818/838]	[2,171/854]
Oil Pumps tested	47	54	108	234
[Approved/ Condemned]	[34/13]	[N.A./N.A.]	[66/42]	[145/N.A.]
Load of Coal checked	151	203	237	89
Loads of Wood checked	34	55	48	43
Packages checked	1,218	1,171	6,389	8,637
Taximeters tested	N.A.	N.A.	N.A.	47
Arrests	52	56	N.A.	132
Convictions	18	41	N.A.	N.A.
Fines	$255	$350	N.A.	$555

Source: All data from Annual Report of the Department of Public Utilities for the years 1912–1915.

Photo 3.10 Adhesive annual inspection decal applied to weighing and measuring devices that pass inspection.

Source: Photograph by the author for Seattle Consumer Affairs Unit. *Seattle Weights and Measures Regulation 1911–2011: A Century of Consumer Protection* (March 26, 2011).

No. 34058] on a measured-mile course on Broadway (now a busy street on Capitol Hill) and for-hire automobile driver license exams (2,648 in 1915).

Washington Supreme Court 'Shrinkage' Case

The Washington State Supreme Court issued a landmark decision in favor of the City of Seattle in the case of *City of Seattle vs Goldsmith* in April 1913. The case concerned the Seattle ordinance requiring that the net contents (weight of product only – excludes tare) be marked on all packaged goods. This decision established a legal precedent that the manufacturer was responsible to ensure that the net contents were accurate at *point-of-sale* even if it was months after the commodity was packaged. This principle was known as '*net contents at time of sale*' and it was the law for the remainder of the century.

> The defendant contended that [packages of raisins, salt and pails of lard] were subject to *shrinkage* and therefore it would be unreasonable and unjust to require the contents to be stated since the

Photo 3.11 Seattle taxicab affiliated with Yellow Cab (c. 1950) (notice Seattle taxicab license on body behind right wheel).

Source: Courtesy of Seattle Municipal Archives [Identifier 191154].

> markings might be incorrect at the time the package reached the consumer although correct at time packed. The Court held that if a commodity is subject to shrinkage no one is in a better position to know that fact than the packer and it is his duty to make allowance for same so that when the package reaches the consumer it shall be full weight. The item of lard was involved in two other cases brought into court at the same time as the Goldsmith case but by agreement it was decided that the one decision should govern the three cases.[13]

Subsequently, in 1913, the U.S. Department of Agriculture redefined *net contents* with a more expansive meaning to allow a tolerance for shrinkage to be deducted.[14] This issue has remained very controversial among weights and measures officials for many years. See Chapter 6 for a full discussion of the legal issues.

In Recent Years

The organization of the Seattle Consumer Affairs Unit remained nearly unchanged for nearly a century although the weights and measures

Photo 3.12 Seattle weights and measures inspector checking a coal wagon (c. 1910).

Source: Courtesy of Seattle Municipal Archives [Identifier 191656].

inspection activities underwent considerable evolution as technology improved. For example, instead of weighing coal wagons making deliveries to households and businesses, inspectors in modern times tested meters used to measure deliveries by home heating oil delivery trucks. Now these trucks are nearly obsolete because of cheaper natural gas furnaces. Scales that had been mechanical devices were now electronic. Taximeters have evolved from mechanical to electronic then to '*virtual*' or GPS.

Farmer's markets were largely replaced by regional and national chains of supermarkets. Bulk sales were replaced by packaged products for nearly everything that consumers buy. As the economy modernized, so have the weights and measures inspections conducted by city weights and measures regulators. Perhaps the most profound changes were driven by the invention of personal computers and their applications in weighing and measuring devices of all kinds.

Photo 3.13 Seattle weights and measures inspector testing meter on back of home heating oil delivery truck (c. 1960). Notice hose delivering heating oil to 300-gallon test measure in the pit below the oil rack. Oil is returned through nozzles on upper deck and top loaded (where man is standing next to safety rails).

Source: Courtesy of Seattle Municipal Archives [Identifier 191683].

Weights and Measures Regulatory Fee Revenue

Some of these changes in the marketplace had profound impacts on the weights and measures regulatory program in Seattle that had not been anticipated. For example, the weights and measures budget had almost been entirely drawn from the General Fund but now the City Council required all regulatory programs to become self-funding through registration (license) fees and inspection fees. Most regulatory fee revenue needed to support the Seattle weights and measures regulatory program was generated by registration fees on commercial weighing and measuring devices. However, the marketplace had largely shifted from bulk sales to package sales. Registration fees were not suitable for goods sold in packaged form. Inspection fees placed an unfair burden on those sellers randomly selected for inspection. Previously, registration fee revenue was simple to project when the inspection plan consisted of '*every device every year.*' Most package inspections were conducted on a '*complaint basis*' due to limited resources so inspection fee revenue was unpredictable and therefore unsuitable for budgeting.

Weights and Measures Inspection Plan

In 1997, the Consumer Affairs Unit *developed* a *strategic plan* to address the cumulative changes in the marketplace during the previous decades due to changing consumer tastes (e.g., packaged products, ready-to-eat food sales in grocery stores) and new technology (e.g., virtual taximeters, water submeters for apartments). This was a long-term plan for consumer protection in a changing marketplace. The goals of the plan were several: "(1) *effectiveness* – reduce industry noncompliance with the provisions of the Weights and Measures Code (Seattle Municipal Code, Chapter 7.04) and the new Taxi Code (SMC, Chapter 6.310); (2) *efficiency* – increase the productivity of inspectors [to reduce costs]; (3) *coverage* – broaden inspection activity to include all weighing and measuring devices; (4) *professionalism* – implement a comprehensive program of inspector training and certification; and (5) *outreach* – expand industry training and consumer awareness initiatives."[15]

The strategic plan was based on the very successful electronic price scanning system inspection program that had started up nearly a decade previously. Price scanning inspection activities are part of the package inspection program. Price scanning inspections were targeted and more frequent at store locations with poor compliance histories. Price scanning compliance training was presented at meetings of managers for chain stores. Inspectors often spot-checked compliance with other package inspection laws (e.g., checking net contents of store-packed meat, package labeling, unit price code compliance, taking tare at checkout and method of sale) *while* already present to conduct price scanning inspections. This made inspectors more efficient. Inspectors attend formal classroom training and participate in on-the-job training before being certified to conduct price scanning inspections. At the time, unannounced price scanning inspections were conducted approximately on an annual basis or more often in response to consumer complaints.

'Report Cards' *on Weights and Measures Inspection Activities*

In order to have a baseline for determining which inspections and locations required more emphasis through accelerated inspections to reduce high levels of noncompliance, a series of Report Cards were prepared for the major areas of inspection. The Report Cards analyzed data from inspection forms. Failure rates alone did not determine which types of inspections would have the most consumer protection impacts given the finite available resources. The number of annual transactions and the size of the overcharges were also important. For example, gas pump inspection failures averaged about 9 percent during 1994–1995 consistently while home heating oil delivery truck meters failed at a rate of 29 percent. However, gas pump errors were minor. And, there were 148 gas stations with 2,431 gas pumps that sold approximately 275,101,515 gallons of gasoline in 1995 at an average price of $1.19 per gallon for regular grade.[16] By comparison, there were only 125 vehicle-tank meters.[17]

Traditionally, since regulatory fees were based on weighing and measuring devices, inspector work was focused overwhelmingly on scales and retail motor-fuel dispensers (gas pumps). The failure rates of other devices (e.g., liquefied petroleum gas meters (propane meters), loading rack meters, vehicle-tank meters) were much higher because they were inspected less frequently. Some devices were inspected by state inspectors, instead of city inspectors since the city lacked specialized test equipment (e.g., vehicle scales, loading-rack meters, LPG meters). Timing devices (parking garages, laundromats) were inspected on complaint basis only. Besides price scanning inspections, virtually no other package inspection activities were conducted before the *Weights and Measures Inspection Plan* was implemented.

Elements of the Weights and Measures Inspection Plan

The elements of the *Weights and Measures Inspection Plan* ('the *Plan*') included: "(1) inspection targeting; (2) variable inspection frequency; (3) streamlined ('two stage') inspection procedures; (4) use of new inspection technology; (5) systematic inspector training and certification; (6) proactive outreach programs for businesses and consumers; (7) new databases and measures of effectiveness (MOE) for evaluating the impact of inspection programs and identifying the need for changes; and (8) increased contacts with media to enhance consumer awareness of weights and measures issues."[18]

Inspection targeting. The *Plan* provides measures of effectiveness (MOE) to quantify the impact of inspections using compliance histories, size of overcharges and number of transactions. The gas pump example previously is instructive. The concept of costs vs benefits applies: inspection costs, opportunity costs (the cost of not conducting the inspection) and the benefit of reduced overcharges.

Inspection frequency. The frequency of inspections will depend upon the compliance history where businesses that pass will have the time between inspections extended. Businesses that fail will be placed on increased frequency. A business must pass two consecutive inspections before being removed from the target list.

Two-stage inspections. Abbreviated inspections, spot-checks or audits, will be used to sample general compliance of an industry. For instance, an inspector will randomly select and test six gas pumps at a 24-pump station. If the sample fails then the inspection will be expanded to all gas pumps. The inspection will also be expanded to all pumps if the average calibration error exceeds -2 cubic inches (known as the 'station average').

Inspection technology. Inspection procedures and technology that reduce time required to conduct inspections will be employed whenever possible. For example, the normal practice of conducting both a *fast fill* and *slow fill* test on every gas pump will be suspended except when an inspection fails. Virtually all consumers use fast fill on a gasoline dispenser nozzle. In addition, use of the '*prover truck*' in lieu of manual inspection with 5-gallon hand provers reduced inspection times by nearly 50 percent. New inspection trucks with service boxes and a hydraulic lift that raises the tailgate above the floor of the service box to the highest loading docks were placed into service. Inside, four-wheeled carts, with stored 25-pound test weights, replaced two-wheeled hand trucks that are manually loaded with 50-pound test weights. Besides saving time, the new equipment reduces lifting injuries. Use of a laptop and off-the-shelf commodity testing software speeds up net contents inspections by automatically performing complex random-testing mathematical calculations (e.g., sample errors).

Inspector training. The federal Office of Weights and Measures in the National Institute of Standards and Technology (NIST) had expanded the number of '*train-the-trainer*' courses that they offer for selected inspectors to become certified trainers. These inspectors, in turn, were expected to make themselves available to travel to jurisdictions in other cities and states to teach training courses for certification of inspectors (travel expenses paid by hosting state). The courses were listed in National Conference on Weights and Measures (NCWM) Publication 20, *NCWM Training Resource Catalog*. Courses were offered on most weighing and measuring device inspections (e.g., small retail computing scales, retail motor-fuel dispensers) as well as some package inspections (e.g., checking the net contents of packaged goods, price scanning). The NIST intended to standardize training for inspectors just as weights and measures laws and inspection standards had been made uniform. Previously, inspector training was all local and primarily limited to on-the-job training. National certifications by NIST improved inspector professionalism and productivity and provided inspectors with heightened credibility when testifying as experts during legal proceedings.

Outreach programs. The Seattle Consumer Affairs Unit undertook a series of steps to educate both the regulated industry and consumers on weights and measures. Since price scanning had expanded quickly in the 1990s and was now in use by nearly all retail stores, ranging from small convenience stores to large box stores and supermarkets, *industry training* began with presentations by inspectors on compliance with new price scanning system requirements. These were often scheduled during periodic store manager meetings. A *video* was produced explaining how inspectors checked the net contents of packaged goods with a demonstration of checking milk container net contents ('gravimetric' procedure converted liquid measure of milk to weight for ease of testing). The video was shown periodically on Seattle's Channel 28. Consumer information about weights and measures was included on the city *website* including an online form for consumer complaints. *Inspection decals* were modified to add a telephone number to make consumer complaints more convenient. The Seattle Consumer Affairs Unit became more active in collaborations with other city, county, regional, state and federal consumer protection agencies and representatives from *print, radio and television media* represented on the local Consumer Protection Roundtable. Regular media releases were prepared on weights and measures activities. Public activities were scheduled to celebrate *National Weights and Measures Week* (first week of March). Finally, the Consumer Affairs Unit prepared two series of flyers to educate consumers and industry: *Fact Sheets* for the public on each of the weights and measures inspection activities in Seattle; and *Information Sheets* on a variety of subjects for the regulated industry to help them comply with applicable weights and measures law. To illustrate, partial lists are shown below:

Fact Sheets

No. 1. *Weights and Measures Inspections in Seattle* (Rev. May 8, 1997)
No. 2. *Taxicab Inspection Program* (Rev. February 4, 1998)
No. 3. *Commodity Inspection Program* (Rev. May 8, 1997)
No. 4. *Gas Pump Inspection Program* (Rev. May 8, 1997)
No. 5. *Scale Inspection Program* (Rev. May 8, 1997)
No. 6. *Oil Truck Meter Testing Program* (Rev. May 8, 1997)
No. 7. *Price Scanning Inspection Program* (Rev. May 15, 1998)
No. 8. *Police Impound Contract* (May 15, 1998)

Information Sheets

No. 1. *Bulk Sales of Wrapped Candy* (January 1, 1997)
No. 2. *Price Scanning Inspections at Small Stores* (February 3, 1997)
No. 3. *Re-Numbering Seattle Taxicabs* (April 17, 1997)
No. 6. *Requirements for Scales* (May 15, 1998)
No. 7. *Inspection of Vehicle-Tank Meters* (July 29, 1997)
No. 8. *Inspection of Length-Measuring Devices* (July 31, 1997)
No. 9. *Method of Sale for Fresh Fruit and Vegetables* (August 6, 1997)
No. 10 *Ready-to-Eat Foods at Grocery Stores* (December 5, 1997)
No. 11. *Pricing of Foods Sold from Bulk by Weight* (February 2, 1998)
No. 12. *Net Contents Declarations* (May 28, 1998)

Databases and reports. Improved electronic recordkeeping was necessary to track inspection histories and calculate *measures of effectiveness* (MOE) in order to implement *the Plan* initiatives for inspection targeting, inspection frequency and two-stage inspections. Off-the-shelf inspection software was not adaptable to these initiatives so simple Access databases were developed. Inspection reports were revised to display *inspection codes* on the reverse thereby making data entry faster and more accurate. *Time in* and *time out* information was added on all inspection reports so that inspector productivity could be tracked. Intracity travel time was not very significant so time per inspection was relatively simple to compute.

Media contacts. Media interest was largely limited to gas pump inspections, price scanning inspections taximeter inspections and taximeter rate changes. Gas pump inspections were of interest to the public because the cost of gasoline was high. Price scanning affected nearly every purchase so there was public interest there as well. The taxicab industry was always quite volatile and regularly in the news. Taxicab service connected every other mode of transportation and was key to the intracity transportation infrastructure. Taxicabs were vital for low income, fixed income and disabled persons who were less likely to own motor vehicles.[19]

There is no question that frequent inspections (typically annual) of scales and gas pumps over many years had effectively improved industry compliance with weights and measures laws (reduced failed inspections). However, there were many other weighing and measuring devices that were largely ignored unless consumers complained. And, the marketplace was changing from bulk sales to package sales creating a need to shift inspection emphasis as well.

The 'Gas Pump Inspection' Plan is shown below to illustrate one area of device inspections contained in the Seattle *Weights and Measures*

Inspection Plan (1997). Four inspectors were assigned to: weighing and measuring devices in the north and south districts (2), price scanning citywide (1) and checking net contents of packaged goods (package inspections) also citywide (1). Every two to three years, inspectors were shifted between assignments. Each inspector scheduled his own inspections pursuant to guidelines in the Plan.

1997 Weights and Measures Inspection Plan Gas Pump Inspections

Inspection Targeting.

- Gas pump locations that fail inspection during the previous 12 months.
- Gas pump locations that are the subject of consumer complaints (24-hour response time)

Inspection Frequency.

- Increased frequency for targeted gas pump locations – every 6 months (vs. 12 months).
- Two-stage inspections for gas pump locations with good compliance histories. Inspect all gas pumps at these gas stations at least once over a three (3) year period.

Inspection Procedures.

- Two-stage inspections: (1) first stage – Handbook 44 inspection of six gas pumps at large gas stations annually,

  (2) second stage – expand inspection to all gas pumps if any of the six gas pumps fails meter calibration tolerance

  (-6 cubic inches) or the average calibration error exceeds the 'station average' (-2 cubic inches)
- Continue fast and slow fill tests on each meter. Do not apply wire-and-lead security seal to adjusting mechanism of meter unless inspector is concerned about possible tampering (e.g., calibrated to minus tolerance vs. zero error).
- Year-round unannounced inspections (vs. seasonal – e.g., good weather).

  Use new paper inspection decals with no effective year dates. Decals now include Consumer Affairs telephone number.

Inspection Technology.

- Use new prover truck for large gas stations and manual 5-gallon test measures (provers) for re-tests of individual gas pumps or at locations with restricted access (e.g., parking garages, boat docks).
- Continue benzene exposure monitoring by Seattle industrial hygienist during high temperature days. Request written report.

Inspector Training.

- Examination Procedure Outline (EPO) for all inspections: EPO#21 "Retail Motor-Fuel Dispensers, Single Product," EPO#22 "Retail Motor-Fuel Dispensers, Blended Product"
- NIST train-the-trainer quota when available: NCWM Course No. 302 "Retail Motor-Fuel Dispensers and Consoles."

Outreach Program.

- Industry – Information Sheet handout for gas stations informing them that locations with poor compliance will be targeted for frequent inspections and other locations will be subject to random two-stage inspections.
- Consumer Awareness – Fact Sheet advising consumers to look for decals on gas pumps and how to make a complaint
- Consumer Awareness – Complete weights and measures video for use on Channel 28.

Recordkeeping.

- Database – Revise database on Access to add data fields for gasoline brand, failure codes and meter calibration errors. Status Report: inspections conducted, failures, rejects. Inspection Planner: gas pumps due in current month by ZIP code.
- Reports – Revise Inspection Report to include 24 lines per page (typical 24-pump station) and change total draw summary (bottom of Report) to show counts for Premium, Mid-Grade, Regular and Diesel.

Relations with Media.

- Consumer Protection Roundtable – Continue participation in monthly meetings. Continue participation in "Consumer Awareness Week" activities.

- Press releases on newsworthy developments (e.g., new prover truck, Annual Report Card).

Specific inspection plans included in *Seattle's Weights and Measures Inspection Plan* (1997) included: Taxicab Inspections, Price Scanning Inspections, Scale Inspections, Gas Pump Inspections, Commodity Inspections, Oil Truck Meter Tests.

Seattle Consumer Affairs Facility (circa 1997)[20]

In 1997, the weights and measures regulatory program administered by the Consumer Affairs Unit (formerly the Weights and Measures Division) operated a small *Metrology Lab* in the old City Hall for calibrating field standards carried by inspectors. The municipal standards were traceable to state standards in Olympia and national standards at the National Institute of Standards and Technology in Washington, D.C. Seattle weights and measures inspectors and administrative staff were assigned to the *Test Station* located at 805 S. Dearborn Street, nearly beneath Interstate 5 and just a few blocks from the Kingdome. The Test Station consisted of a single-story office building (formerly a gas station that had been rebuilt) with an attached drive-thru taxi bay for testing taximeters on a *dynamometer* (free-spinning rollers for rear drive wheels to simulate a measured mile course). A small customer waiting room was used for licensing functions – primarily taxicab vehicles, taxicab drivers and taxicab associations. In front of the building was a two-story loading rack with a roof for testing vehicle-tank meters installed on home heating oil delivery trucks. Beneath the loading rack was a pit with three stationary test measures with capacities of 300-gallons, 500-gallons and 1,000-gallons. Sight glasses on the vertical fill pipes atop the test measures indicated the error in cubic inches (231 cubic inches equals 1 U.S. gallon). There were about 125 vehicle-tank meters on home heating-oil delivery trucks currently in service.

Seattle Consumer Affairs Vehicle Fleet

The Consumer Affairs Unit leased special-purpose vehicles from the Fleets Department: (1) three Ford F-250 chassis with tall service boxes and rear hydraulic wide liftgates were inspector trucks; (2) a 'prover truck' with three mounted 5-gallon test measures (provers) that drained into individual internal independent tanks and a 50-gallon test measure for diesel fuel dispensers (dispense at a higher rate); (3) an unmarked sedan with an installed taximeter for enforcement activities (e.g., overcharges) associated with the taxicab industry; and (4) two city sedans for weights and measures enforcement (e.g., price scanning inspections, checking net contents of packaged goods). The service boxes on the inspector trucks housed a 4-wheel test cart loaded with twenty 25-pound hand-weights. The test cart wheels were in tracks welded to the floor of the service box. The tracks were continuous to stops on the hydraulic liftgate. The liftgates were designed to be raised higher than the floor of the service boxes to the highest loading docks. An electric winch allowed the cart to move along the tracks and secured in a housing at the front of the box. This arrangement allowed inspectors to move test weights without lifting them. Inspectors avoided the need to make two lifts of 500 pounds from the box floor to the lift gate and then onto the loading dock. These field standards were added in increasing load tests designed to check errors under various loaded conditions on medium and large capacity dormant scales commonly located in warehouses or industrial facilities.

Seattle Consumer Affairs Staff

The staff consisted of a manager, five licenses and standards inspectors (weights and measures – 4, taxicabs – 1) and one license specialist. By 1997, Seattle was the only remaining municipal weights and measures regulatory program. The programs in Tacoma, Everett and Spokane had closed due to budgetary considerations. The Washington weights and measures regulatory program consisted of about 12 inspectors who worked out of their homes from locations scattered all over the state. The state weights and measures program manager and administrative staff were located in the Washington Department of Agriculture offices in Olympia. The state program was funded by device registration fees. Few package inspections were conducted because the State Attorney General's Office had determined that it wasn't legal to use dedicated regulatory fees from the weighing and measuring device inspection program to subsidize package inspections.

By 2015,[21] both the duties and staff of the Consumer Affairs Unit would greatly expand. Three new regulatory programs were added: limousine industry, towing industry and transportation network company (TNC) industry (e.g., Uber, Lyft). The facility was expanded with a portable office trailer (five inspectors) and the overflow (six inspectors) were housed in a former training room in another building at the Dearborn Street city facility. The increased staff consisted of 22 positions: one Manager, 13 Licenses and Standards Inspectors (weights and measures, taxicab and TNC inspectors), six Administrative Specialist II (license specialists), one Research and Evaluation Assistant II (database management), and one Office Aide. Most of the expanded duties were related to transportation regulation and not weights and measures. Three-member inspector teams conducting street enforcement worked rotating shifts and random days.

Commercial Weighing and Measuring Devices in Seattle (1997)

The primary weights and measures inspection workload conducted in Seattle during 1997, the first year that the *Weights and Measures Inspection Plan* was implemented, are described below:

Scales. There were approximately 2,500 commercial scales of all kinds in the city of Seattle ranging from small 30-pound capacity computing scales found in many retail applications, such as grocery stores, to large permanently installed dormant scales with capacities in excess of 1,000 pounds commonly used by manufacturing, wholesale and moving and storage businesses. These counts do not include vehicle scales or railroad scales which required expensive special test equipment that only the state weights and measures program possessed. The same applied to liquefied petroleum gas meters (e.g., LPG or propane) and loading rack meters (used to load gasoline tank trucks that supply gas stations). The number of scale inspections conducted in 1995 were: small capacity scales (<30 pounds) – 2,266, medium capacity scales (31–1,000 pounds) – 487, and large scales (>1,000 pounds) – 170. Many of the small electronic scales used in retail '*point-of-sale*' applications (e.g., grocery store check-out lanes) have standard '*tare*' (i.e., packaging) values programmed so that consumers are only charged for the '*net contents*' of the package. The failure rate of small scales was about 5 percent and errors were usually minor. Large supermarkets would often have 20 or more small scales – at all check-outs, most departments (e.g., meat department, deli counter, produce department) and at '*ready-to-eat*' food lines.

Gas Pump Meters. There were approximately 3,000 gasoline and diesel pumps (retail motor-fuel dispensers) located at 150 gas stations in Seattle. There were a few located in parking garages or at marinas but most were at 24-pump gas stations. Maintenance tolerances for annual inspections were plus or minus 6 cubic inches or less than 3/100ths of a gallon in a 5-gallon test measure. There are 231 cubic inches to 1 U.S. gallon and 1 cubic inch equals 0.004329 gallons. An individual gas pump meter fails if the error exceeds -6 cubic inches. The entire gas station fails if the *average* error exceeds -2 cubic inches (i.e., short measure). The latter requirement was to prevent a service company from calibrating gas pumps to average the maximum negative error allowed (-6 cubic inches). Blending pumps combined two grades (e.g., regular and premium) to produce a mid-grade octane so there are three grades but just two meters. Counting devices for the purpose of collecting registration fees was complicated by blenders. A special prover truck with three installed 5-gallon test measures on gimbals (to level sight glasses) was employed for testing different grades of gasoline. Each test measure displayed errors with a sight glass. After each test, gasoline was drained into three integral (built-in) tanks, which were then gravity-drained to return the gasoline to the underground storage tank for that grade at the gas station. Inspectors took fuel samples to a designated contract lab for analysis when they received water contamination or low octane complaints from consumers. Gas station operators are supposed to sound storage tanks manually with sticks coated in *water paste* (changes color in presence of water that collects at the bottom of underground storage tanks) to monitor water collection and pump it out.

Vehicle-Tank Meters. The number of home heating oil delivery tank-trucks with installed meters (vehicle-tank meters) has been on the decline for many years as households replace oil furnaces and buried storage tanks with electric or natural gas furnaces. During 1997, 125 vehicle-tank meters were tested at the Test Station with a failure rate of 29 percent. The procedures for conducting these tests were outlined in Seattle Information Sheet No. 7 “Inspection of Vehicle-Tank Meters” (July 29, 1997) that was distributed to the industry. These instructions were based on the national standard, Section 4.31 “Vehicle-Tank Meters,” contained in National Institute of Standards and Technology (NIST) Handbook 44, *Specifications Tolerances, and Other Technical Requirements for Weighing and Measuring Devices*. Inspections are conducted in accordance with, National Institute of Standards and Technology (NIST) Publication 12, *Examination Procedure Outlines for Weighing and Measuring Devices*, EPO No. 23 “Vehicle-Tank Meters, Power Operated.” Normally, the 300-gallon test measure located in the ‘pit’ below the loading rack at the Test Station was used to check compliance with *maintenance* tolerances (tolerances for devices already in service). The tests were conducted at the maximum discharge rate. The maintenance tolerance was -175 cubic inches (approximately -0.8 gallon). The sight glass on the neck of the 300-gallon test measure indicates the error, minus (short measure) or plus (giving away product). Meters were not failed for positive errors (i.e., giving away product). There is a ‘*split-compartment*’ test to ascertain whether the air eliminator properly prevents the meter from recording the passage of air when suction is switched from an empty oil compartment to another compartment with product. Interestingly, a sizeable number of delivery truck drivers worked on fishing vessels in Alaska during Summer months since heating oil deliveries were seasonal (usually Fall/Winter).

Taximeters. Installed taximeters were tested annually in the Taxi Bay of the Test Station. They are measuring devices although taxicab service is sold by distance and time instead of mass or volume. The taximeter fare is the combination of the '*drop charge*' (first increment of distance, e.g., $1.80 for first 1/9 mile), '*distance charge*' ($0.20 for each additional increment of distance, e.g., second and subsequent 1/9-mile increments), and '*time charge*' (each increment of time, e.g., $0.20 for each 24 seconds or $0.50 per minute). This taximeter rate was in effect in 1997. All charges are assessed in advance or at the start of a new increment of distance or time and there are no prorated charges. For the test, taxicabs pull onto a dynamometer in the Taxi Bay. The dynamometer has a counter that records revolutions as the taxicab drive wheels spin on a pair of steel rollers. Distance travelled on a '*simulated measured mile*' test is computed by multiplying revolutions by the circumference of the steel rollers. In this case, the circumference is 2.833 feet so 1,864 revolutions equal one mile. The taximeter test consists of nine '*drops*' or events where the total fare displayed increases ($1.80, $2.00, $2.20 …). The driver and inspector are seated in the taxicab during the test. The driver accelerates to approximately 20 miles per hour for the test (a speed above the crossover speed). The '*crossover speed*' is that speed where the distance charge equals the time charge. The crossover speed is calculated by dividing the hourly time charge by the distance charge: $30.00 per hour/$1.80 per mile = 17 mph. When a taxicab slows to a speed less than the crossover speed, the taximeter charges by time instead of distance. The taximeter continues to add time charges while stopped at a traffic light or stop sign. A taxicab driver pays a lease for a shift and would not earn any fare revenue to cover that expense unless the taximeter continued to add time charges at a stop light. The tolerance on over-registration (overcharging) is 1 percent against the customer. Once a taximeter test is passed, a wire-and-lead security seal is affixed to the calibration mechanism to prevent tampering. The Seattle Consumer Affairs Unit tested taximeters for King County taxicabs, taxicabs licensed by the City of Everett and taxicabs licensed by the City of Seattle under cooperative agreements between the governments. The first taximeters were mechanical and directly connected to the transmissions of the taxicabs. Electronic taximeters replaced these and were generally connected to the cruise control box in the engine compartment. Most recently, taximeters use GPS to measure distance travelled. Transportation network companies, such as Uber and Lyft, developed these '*virtual taximeters*,'[22] which were approved for use after testing under the California Type Evaluation Program. There is a temporary national standard for GPS taximeters in NIST Handbook 44 but no NTEP (National Type Evaluation Program) *Certificate of Conformance* (CC) is available.

Price Scanning Inspections. In 1997, the Seattle Consumer Affairs Unit conducted 141 price scanning inspections including 69 100-item or 'full' inspections. Inspections at convenience stores or other small retail stores were limited to 25 or 50 items. Inspections at supermarkets and big box stores consisted of verification of prices for 100 items. About 36.2 percent of inspections had zero pricing errors against the customer (overcharges) and failed inspections (more than 2% overcharges) were 14.4 percent. There was an average of 3.2 percent total pricing errors comprised of 1.5 percent overcharges and 1.7 percent undercharges. A national study conducted in 1996 by the Federal Trade Commission (FTC) and the National Institute of Standards and Technology (NIST) found that pricing errors were higher in states surveyed compared with Seattle because Seattle already had an active price scanning inspection program for more than a decade. See Table 3.2.

The national testing standard, "Examination Procedure for Price Verification," is contained in NIST Handbook 130 *Uniform Laws and Regulations*. It provides inspection procedures and enforcement practices. In law, violations of "Misrepresentation of Price," at Section 7.04.505 of the Seattle Municipal Code could result in a $200 fine or 3 months imprisonment (Sec. 7.04. 690.B) but, in practice, stores were placed on accelerated inspection frequency until they passed two consecutive inspections. The emphasis was on voluntary compliance instead of enforcement.

Commodity Inspections. The Seattle Consumer Affairs Unit had one full-time inspector dedicated to checking the net contents of packaged goods. Daily inspections of the net contents of packages were conducted both at retail and '*point-of-pack*.' These were '*standard packs*' or packages labeled with the same net contents (e.g., 1 gallon of milk) and

Table 3.2 Comparison of Price Scanning Errors 1996 FTC/NIST Study and Seattle Inspections

1996	*States Surveyed by FTC/NIST*	*Seattle*
Total Errors	4.82%	2.8%
Undercharges	2.58%	1.8%
Overcharges	2.24%	1.0%

'*random packs*' with differing net contents (e.g., store-packed meat). The test procedures are contained in NIST Handbook 133 *Checking the Net Contents of Packaged Goods*. Typically, an inspection sample consisted of a random sample of packages selected, using a random number generator, from the *population* present at the retail store, warehouse (wholesale) or the manufacturing facility (production). The average tare for the sample is determined then entered into the inspector's portable electronic scale. Next the gross weights of the sample packages were weighed to ascertain whether each is over or under the target weight (net weight of contents plus tare weight). No individual package can be short weight in excess of a specified '*Maximum Allowable Variation*' (MAV) and the average of the entire sample must equal or exceed the *target weight*. If the sample fails, for MAV or *average*, then the entire population fails and must be relabeled with an accurate net contents declaration before it can be sold. For very large samples, there may be more than one MAV permitted. For packages containing liquid product, the liquid measure is converted to a weight in a procedure referred to '*gravimetric testing*.' For example, the net weight of one gallon of milk would be verified by gravimetric testing. The contents of the package would be poured into calibrated glassware until the bottom of the meniscus of the milk sample rested on the mark for the one-gallon test measure then it was weighed. That weight, less the weight of the empty test measure, is the net weight of one gallon of milk. The weight will vary depending on whether it is whole milk, 1 percent, or 2 percent and whether it is white or chocolate and the date of the production run. At the start of each production run, sample packages are weighed using a '*check weigher*' to make certain that all packages meet or exceed the *target weight* for the product. The speed of the bottling line can result in '*short fill*' if containers are filled too fast and the contents foam. Checking net contents is done at dairies, soft drink bottling plants, coffee roasters, bakeries, drywall joint compound plants as well as retail stores that offer these products for sale to the public. Special net contents procedures and test equipment is required for certain products. For instance, bags of mulch or '*designer dirt*' (potting soil) settle when dumped into a test measure to verify volume (e.g., cubic feet) so it must be slowly and gently poured. Pre-mixed drywall joint compound is packaged in plastic bags of irregular shape then packaged in boxes. It is sold by liquid measure as an exception to the general rule that solids and semi-solids be sold by weight. The plastic bag of product is lowered into a bath of water filled to a spout and the amount of water displaced is weighed. Because of the need to identify a population and then select a random sample, a considerable amount of product handling is required. Also, there are many steps to the net contents procedure involving numerous mathematical computations to account for the sampling errors inherent in small samples. This makes the process of checking net contents very time consuming. A typical inspection generally requires about two to three hours to complete.

Current Weights and Measures Inspections in Seattle

The weights and measures program in Seattle declined during the past 20 years. There were economic downturns that prompted the City Council to move budgets from out of the General Fund to regulatory fees – a form of self-funding. The cost of living in Seattle was considerably higher than rural areas or smaller cities so caps on weighing and measuring device registration (license) fees, that had been set low by the state legislature under pressure from industry lobbyists, proved inadequate to support inspector salaries in Seattle. Washington weights and measures inspector salaries were lower because they were assigned to small cities and rural areas around the state with significantly lower costs of living. The Washington weights and measures program was unable to introduce a bill to the state legislature to institute a price scanning system registration fee similar to Seattle so they don't have funding to support price scanning system inspections. The State Attorney General has issued an opinion that it is not proper to use registration fee revenue from weighing and measuring devices to subsidize price scanning system inspections. All regulatory fee revenue, from device registration fees and inspection fees, must be spent to support the program where it was raised. In Seattle, across-the-board budget cuts during recessions led to the loss of one inspector position (leaving three full-time inspectors) and the loss of two inspector trucks (leaving one inspector truck, one prover truck and a sedan). This amounted to a 25 percent reduction in resources and had a big impact on inspection activities. As a result, inspection frequencies were lengthened and '*audits*' (partial inspections) were substituted for full device inspections and net contents inspections. '*Two-stage*' inspections were adopted – an audit consisting of an abbreviated inspection then a full inspection if the audit failed. Enforcement action required that full NIST Handbook inspection procedures be followed. For example, an audit of retail motor-fuel dispensers meant that the 'slow fill' tests were discontinued and just six of 24 gasoline meters were tested. In addition, service stations with histories of failed inspections were targeted for more frequent inspections. Consumer complaints were still the highest priority and prompted immediate inspections. The weights and measures program in Seattle still attempted to maintain a balance between device and package inspections. Most weights and measures regulatory programs in the United States were faced with the same resource challenges and they responded similarly.

Notes

1 *Proceedings of the Annual Convention of the League of Washington Municipalities* at Spokane, Washington November 19–22, 1913. Sixth Session: "Department of Weights and Measures" by A. W. Rinehart November 21, 1913, p. 94.

2 Arthur W. Rinehart was selected by the Washington Secretary of State, I. M. Howell, to become the first Chief Inspector of the new State Department of Weights and Measures on May 31, 1913. "Rinehart leaves Service of the City" *The Seattle Sun*, May 31, 1913. Cited in, Seattle Consumer Affairs Unit. *Seattle Weights and Measures Regulation 1911–2011: A Century of Consumer Protection* (March 26, 2011), p. 2.

3 "Rinehart Appointed Weights Inspector" *The Seattle Sunday Times*, April 23, 1911.

4 "Leslie J. Allen Gets City Appointment" *The Seattle Sunday Times*, May 7, 1911.

5 "Promotions Follow Rinehart's Elevation" *The Seattle Times*, May 31, 1913.

Craig Leisy. "Then and Now: Seattle Weights and Measures" *Xchange* (Executive Services Department employee newsletter), p.1. Leslie J. Allen served as Chief Inspector of the Seattle Weights and Measures Division in the Department of Public Utilities for 27 years until he retired on May 30, 1940.

Seattle Consumer Affairs Unit. "Now and Then…" City of Seattle web site [obsolete] www.cityof seattle.net/elizabeth/consumer/history.htm December 4, 2000.

6 "New City Official Assumes Charge" *The Seattle Post-Intelligencer* April 22, 1911.

7 City of Seattle. *Annual Report of the Department of Public Utilities, 1913*, p. 95.

8 City of Seattle. *Annual Report of the Department of Public Utilities, 1912*, p. 95.

9 Ibid.

10 Ibid., p. 101.

11 Seattle Consumer Affairs Unit. "Now and Then…" City of Seattle web site [obsolete] www4.cityof seattle.net/elizabeth/consumer/history.htm December 4, 2000.

12 "War on Short Milk Measure" *Seattle Post-Intelligencer*, May 27, 1912.

13 City of Seattle. *Annual Report of the Department of Public Utilities, 1913*, p. 96. Also, see *Proceedings of the Annual Convention of the League of Washington Municipalities* at Spokane, Washington November 19–22, 1913.

Sixth Session: "Department of Weights and Measures" by A. W. Rinehart November 21, 1913, pp. 95–97.

14 A 1913 amendment to the to the Pure Food Law stated, in part, "Provided however, that the secretary of commerce, secretary of agriculture and the secretary of the treasury shall make tolerances and allow variations for the marking of packaged goods." *Proceedings of the Annual Convention of the League of Washington Municipalities* at Spokane, Washington November 19–22, 1913. Sixth Session: "Department of Weights and Measures" by A. W. Rinehart November 21, 1913, p. 95.

15 City of Seattle, Consumer Affairs Unit. *Weights and Measures Inspection Plan* (March, 21, 1997), p. 1.

16 City of Seattle, Consumer Affairs Unit. *Gas Pump Inspection Report Card* (1995), September 23, 1996, p.1, N. 1.

"The population of Seattle was 516,259 in 1990 (U.S. Census). The same year, the number of motor vehicles in Washington state was 875 per 1,000 resident population and average fuel use was 609 gallons per vehicle (U.S. DOT *Selected Highway Statistics and Charts* (1990), p. 31). As a result, approximately 275,101,515 gallons of motor fuel are consumed each year in Seattle. The typical price per gallon in 1995 was $1.19 for Regular and $1.38 for Premium (Federal Highway Administration Monthly Motor Fuel Reported by States (Table MF-5) February 1995)."

17 City of Seattle. Consumer Affairs Unit. "Oil truck Meter Testing Program" Fact Sheet No. 6 (Rev. May 8, 1997).

18 City of Seattle, Consumer Affairs Unit. *Weights and Measures Inspection Plan* (March, 21, 1997), p. 2.

19 Craig A. Leisy. *Transportation Network Companies and Taxis: The Case of Seattle*. London: Routledge (2019). This volume contains an economic history of the taxicab industry which started up nearly concurrent with the Seattle weights and measures program. Because taximeter rates were set by the City of Seattle and taximeters were measuring devices regulated by the weights and measures program, the Consumer Affairs Unit in Seattle regulated both weights and measures and the taxicab industry. Later, the Consumer Affairs Unit in Seattle also commenced regulating transportation network companies (e.g., Uber, Lyft), limousines and the towing industry (police impounds).

20 The author became the Supervisor (later, the Manager) of the Seattle Consumer Affairs Unit in early January 1996 and retired in September, 2017 (22 years). He had in-depth, first-hand knowledge of all aspects of the weights and measures program throughout. He was an active member of the National Conference on Weights and Measures and was appointed to chair a NCWM working group that prepared a landmark study, *Survey of Inspection Statistics Collected by State Weights and Measures Programs [2002]*. This was the first study of its kind and it documented the lack of uniformity among programs in different jurisdictions which has guided much of the work of the NCWM in subsequent years.

21 Three new regulatory programs were undertaken by the Consumer Affairs Unit in the preceding years: limousine industry regulation (2012), towing companies (2013) and transportation network companies, e.g., Lyft, Uber (2014).

22 The Consumer Affairs Unit proposed to NCWM that the Taxi Code in NIST Handbook 44 be amended to provide a new standard for 'virtual taximeters' or fare calculation using GPS. For more information, see: Craig A. Leisy. *Transportation Network Companies and Taxis: The Case of Seattle* New York: Routledge (2019), p. 8.

Bibliography

City of Seattle. *Annual Report of the Department of Public Utilities, 1912.*

City of Seattle. *Annual Report of the Department of Public Utilities, 1913.*

Leisy, Craig A. *Transportation Network Companies and Taxis: The Case of Seattle*. London: Routledge (2019)

League of Washington Municipalities. *Proceedings of the Annual Convention of the League of Washington Municipalities* at Spokane, Washington, November 19–22, 1913.

4 The National Institute of Standards and Technology (NIST) and National Conference on Weights and Measures (NCWM)

This chapter will discuss the purpose, organization and history of the National Institute of Standards and Technology/Office of Weights and Measures (NIST/OWM) and the National Conference on Weights and Measures (NCWM). In addition, the role of NIST and NCWM with regard to state and local government regulatory programs will be examined.

The National Bureau of Standards

The National Bureau of Standards (NBS) was established by an act of Congress on March 3, 1901 (31 Stat. 1449).

> Section 1. *Be it enacted by the Senate and House of Representatives of the United States of America in Congress assembled*, That the Office of Standard Weights and Measures shall hereafter be known as the National Bureau of Standards;
>
> Section 2. That the functions of the bureau [forerunner of the National Institute of Standards and Technology (NIST)] shall consist in the custody of the standards; the comparison of the standards used in scientific investigations, engineering, manufacturing, commerce and educational institutions with the standards adopted or recognized by the Government; the construction, when necessary, of standards, their multiples and subdivisions; the testing and calibration of standard measuring apparatus; the solution of problems which arise in connection with standards; the determination of physical constants and the properties of materials, when such data are of such great importance to scientific or manufacturing interests and are not to be obtained of sufficient accuracy elsewhere.[1]

DOI: 10.4324/9781003263661-5

First Conference of State Sealers of Weights and Measures

The National Bureau of Standards invited representatives from all states to attend a meeting of the sealers of weights and measures of the United States in Washington, D.C. during April 1904.

> February 13, 1904
>
> DEAR SIR: In order to bring about uniformity in the State laws referring to weights and measures, and also to effect a close cooperation between the State inspection services and the National Bureau of Standards, it is proposed that a meeting of the State sealers of weights and measures (or custodian of the State standards, if there be no sealer) be held in Washington the coming spring. It is our opinion that such a meeting would afford an opportunity for exchange of views and for the discussion of the questions involved, and would lead to a better solution than could be obtained in any other manner.
>
> In case it is finally decided to hold such a conference, would your State send a representative; and if so, would April 15 be agreeable to him?[2]

There wasn't adequate time to make sufficient arrangements for the meeting so it was rescheduled for January 16–17, 1905 instead of April 15, 1904. Delegates from Iowa, Kentucky, Massachusetts, Michigan, New Hampshire, Pennsylvania, Vermont, Virginia and the District of Columbia attended. Also present were Mr. S. W. Stratton, Director of the National Bureau of Standards and Mr. Louis A. Fischer, Chief of the Weights and Measures Division.

> The first conference, meeting in January 1905, with representatives from seven States and the District of Columbia, disclosed that in most of these States the laws relating to weights and measures were "exceedingly lax…with nothing obligatory" or were "practically a dead letter", that the State sealer's office was usually unsalaried, and the duties of county sealers were often imposed on the county treasurer or even the superintendent of school. In more than one State, the county and city sealers were not compelled to procure standards, and several of the State representatives knew nothing about their State standards or even where they were to be found.[3]

Second Conference of State Sealers of Weights and Measures

A second conference was convened at the Bureau of Standards in Washington, D.C. during April 12–13, 1906. A letter was sent by S. W. Stratton, Director of the Bureau of Standards, to the governors of the States.

> DEAR SIR: The second annual meeting of the State sealers of weights and measures will be held in Washington,
>
> D.C. on April 12, 1906, and it is earnestly requested that your State be represented. The object of these meetings, as stated in former communications to you, is to improve conditions affecting commercial weights and measures.
>
> The functions of the Bureau of Standards include the construction and verification of State standard weights and measures, but the use of these standards for the regulation of commercial weights and measures is a function which has been largely left to the State and municipal authorities, and in the exercise of which the Bureau of Standards is ready to assist.
>
> It is evident from the number of convictions for the use of dishonest weights and measures in localities where rigid inspection is maintained that the amount of fraud in State and cities where there is no inspection, or inefficient inspection, must be enormous; and, unfortunately, the loss usually falls upon those too poor or unintelligent to protect themselves.
>
> The Bureau was led to take up this matter by the increasing number of inquiries received from the citizens of every State in regard to weights and measures matters which could only be properly attended to by local inspectors.
>
> It is believed that the free interchange of views and experience will result in the passage of a Federal law applicable to all the States, or in the enactment of uniform laws by the separate States.
>
> At the first meeting, held at the Bureau on January 16 and 17, 1905, facts were brought out which showed that almost all the States have laws concerning standards and commercial weights and measures, but only a few have made the necessary provisions for their enforcement. An examination of the proceedings of the first meeting, a copy of which is forwarded to you under separate cover, will show how little has been done in this direction and how much room there is for improvement.
>
> It is earnestly requested that you inform the Bureau of what action is taken in the matter, and if a delegate is appointed that his

> address be furnished in order that he may be supplied with whatever literature we have on the subject of weights and measures.[4]

Delegates from Arkansas, California, Connecticut, Delaware, Florida, Kansas, Kentucky, Maryland, Massachusetts, Montana, New Hampshire, New Mexico, New York, Ohio, Pennsylvania, Rhode Island, Vermont, Virginia, Wisconsin, Wyoming and Washington, D.C. attended the second conference. Also present were Mr. S. W. Stratton, Director of the National Bureau of Standards, and Mr. Louis A. Fischer, Chief of the Weights and Measures Division.

> At the second conference, in April 1906, it was decided to set up a permanent organization of State officials, make the conference an annual event to discuss the testing and sealing of commercial weights and measures, and work toward adoption of uniform laws.[5]

Third Conference of State Sealers of Weights and Measures

Delegates from Colorado, Georgia, Illinois, Iowa, Kansas, Maryland, Massachusetts, Michigan, Montana, New York, Ohio, Pennsylvania, Rhode Island, Virginia, West Virginia and Washington, D.C. attended the third conference. Also present were Mr. E. B. Rosa, Acting Director of the National Bureau of Standards, and Mr. Louis A. Fischer, Chief of the Weights and Measures Division.

> Seventeen States were represented at the third conference in 1907 and as at the previous meetings the discussion soon centered around "the question of honest weights and measures in all business transactions", the almost infinite variety of laws affecting weights and measures, and the meager funds provided by the States for their inspection. The conference began work on a model weights and measures law, to be offered for adoption of all the States, and recommended unanimously that additional powers be given the Bureau of Standards to make the State laws effective.

During the third conference, a model weights and measures law proposed by the attendees was presented. There were 34 sections that addressed national law, local law and general regulations.

> SUGGESTIONS FOR NATIONAL AND STATE LAWS ADOPTED BY THE NATIONAL CONFERENCE ON

WEIGHTS AND MEASURES AT THE THIRD ANNUAL CONFERENCE, MAY 16–17, 1907

NATIONAL LAW

...

Sec. 4. No weighing or measuring device shall be used for the purpose of trade until the type has been approved by the National Bureau of Standards. Any type so approved may be used anywhere in the United States [*eventually: National Type Evaluation Program*]: *Provided*, That nothing in this act shall prevent the state commissioner of weights and measures or local inspector from condemning such device if its operation should be defective

Sec. 5. Model regulations for the guidance of state commissioners of weights and measures and local inspectors shall be prepared by the National Bureau of Standards in cooperation with the National Association of State Commissioners [*later: National Conference on Weights and Measures*].

Sec. 6. The model regulations, prepared and issued by the National Bureau of Standards, shall govern the procedure to be followed by the state commissioner of weights and measures and local inspectors in inspecting, testing, and sealing all weights, measures, balances or measuring devices [*eventually: NIST Handbook 44, Handbook 130*].

Sec. 7. The net quantity of the contents of all packages shall be plainly stated in terms of weight or measure on the outside of the package [*eventually: labeling, checking net contents of packaged goods*].

LOCAL LAW

...

Sec. 14. Every county and municipality in the State shall appoint a sealer, with a sufficient number of deputies to inspect at least once a year every weight, measure, balance or measuring device of any kind used in trade within the jurisdiction of said county or municipality [*presumed source of annual device inspections*].

GENERAL REGULATIONS

...

Sec. 27. No weights, measure, balance, or measuring device of any kind shall be used in trade until it has been examined and sealed by the state commissioner of weights and measures or local sealer.[6]

Eventually, at the tenth annual conference (1915), the national standard, *Tolerances and Specifications for Weights and Measures and Weighing and Measuring Devices*, was adopted by the National Conference on Weights and Measures (NCWM) and published by the National Bureau of Standards (NBS).[7] It was the forerunner of NIST Handbook 44, *Specifications, Tolerances, and other Technical Requirements for Weighing and Measuring Devices* (2020).

Fourth Conference of State Sealers of Weights and Measures

The fourth annual conference was held at the National Bureau of Standards in 1908. Delegates from Colorado, Illinois, Iowa, Kansas, Kentucky, Massachusetts, New York, Ohio, Pennsylvania, Rhode Island, Vermont and Washington, D.C. attended the fourth conference. Also present were Mr. S. W. Stratton, Director of the National Bureau of Standards (NBS), and Mr. Louis A. Fischer, Chief of the Division of Weights and Measures.

During the fourth conference, the question of national (federal) weights and measures legislation versus *state adoption of uniform weights and measures laws* was discussed further. Eventually, state adoption became the method of attempting to achieve uniform weights and measures regulation. Early discussions of the issue of '*shrinkage*' (later: 'moisture loss') were also discussed primarily about flour and butter.[8]

Survey of State Enforcement by National Bureau of Standards

States were slow to properly fund, staff or equip their weights and measures programs. As a result, the NBS sought and received funding to conduct a survey of State use of standards distributed to them for weights and measures regulation.

> The annual conferences of State sealers at the Bureau made it clear that through ignorance and neglect of State responsibilities the American public was being robbed of enormous sums daily in the marketplace. Since the State governments showed little interest in weights and measures reforms, said Stratton, the Bureau "must reach the public through State and city officials by testing their standards." In December 1908 he asked Congress for a special grant of $10,000 "to investigate what the States are doing with their standards and to encourage them to take up and supervise the local work as they should," It was the Bureau's first request for special funds, and Congress approved it without question. What Stratton intended was an investigation to reveal the extent of false and fraudulent weights and measures in use throughout the Nation.

> Between 1909 and 1911, inspectors from the Bureau visited every State of the Union, testing over 30,000 scales, weights, and dry and liquid measures in 3,220 different shops and stores. They were not surprised to find that almost half the scales tested were badly inaccurate, that 20 percent of the weights, half the dry measures, and a quarter of the liquid measures favored the storekeeper [errors of over-registration viz., resulting in overcharges]. The Bureau estimated that in the case of pint butter alone the annual loss to the consumer, through rigged or faulty weighing devices, amounted to more than $8,250,000.[9]

There followed a gradual public outcry for accurate weights and measures after the survey results were known. Consequently, the States began adopting into law the *model law* developed by the NBS and reviewed during the annual conferences of the State sealers. The current status of State adoption of model laws and technical standards is reported annually in NIST Handbook 130 in a Table, "Uniformity of Laws and Regulations." A full discussion of this subject is provided later in this chapter.

Annual Meetings of the NCWM

> Ninety-eight officials representing 25 States and 34 cities attended the Bureau conference held in February 1912, and except during wartime years, these conferences have been held annually ever since.[10]

Mr. S.W. Stratton, the Director of the National Bureau of Standards, served as the first chairman of the National Conference on Weights and Measures for the first fifteen annual conferences (1905–1922). State chairmen were elected beginning in 1958 (43rd conference) and thereafter. No annual conferences were held in 1909, 1917–1918 (WWI), 1932–1934 (Great Depression), 1942–1945 (WWII), 1948 or 2020 (COVID-19 pandemic).

The 104th National Conference on Weights and Measures was held at Milwaukee, Wisconsin during July 14–18, 2019. The 105th conference was not held in 2020 due to the COVID-19 pandemic and was rescheduled online during January 10–12, 2021. The NCWM convenes the annual conferences now instead of the National Institute of Standards and Technology (replaced the National Bureau of Standards). The proceedings of the conferences are published and distributed by NIST/OWM in a series of special publications.

The National Conference on Weights and Measures

The early beginnings of the National Conference on Weights and Measures (NCWM) were coincident with the second conference in 1906. Model laws were a product of the third conference in 1907. Subsequently, the NCWM evolved into the current organization.

> The NCWM is an organization of state and local weights and measures officials, and representatives of industry, consumers, and federal agencies that work together to develop uniform weights and measures laws and regulations, which are published as NIST Handbooks 44, 130 and 133.[11]

More generally, NCWM is a standard-setting organization whose members are regulatory officials that propose and vote on changes to model laws and regulations and technical standards for weighing and measuring devices. Proposed amendments are submitted to one of the four regional weights and measures associations where they are vetted then passed along to the NCWM with recommendations during annual conferences. The NCWM considers voting items and, if they are passed, the NIST Handbook (series) is amended accordingly. Representatives from the regulated industries are considered associate members. They are non-voting members but they participate on the standing committees.

The *NCWM Task Force on Planning for the 21st Century* issued a Final Report in 1992 and issued an opinion on whether voting rights for associate members should be expanded.

> The Task Force debated the issue of whether other membership classes than active weights and measures officials should have a vote within the Conference. The industry representative stated that industry representatives wanted a vote within the Conference.
>
> The standards developed by the Conference are intended to be adopted by State and local weights and measures jurisdictions. It has been the long-standing tradition that State and local government representatives needed the reassurance that the national standards developed by the Conference were the final decision of government officials like them, not of industry. The number of industry members exceeds the number of active weights and measures officials in the Conference; interim and annual meeting attendance is often heavily skewed toward industry representation. The trust developed over the years in the standards would be in jeopardy if industry were given the vote on the base standards developed by the Conference,

> for example, on Handbooks 44 or 130 … The interrelationship and partnership of government and industry within the Conference is most effective and unusual in spite of industry not having a vote on the base standards. If industry representatives are considered for voting status, a balance with consumer interests would become necessary.[12]

Generally, the expertise of manufacturers of weighing and measuring devices is important in developing standards that reflect the latest technology. However, there would be a perceived conflict of interest in allowing industry to vote on the very standards that are used to regulate them.

The NCWM Constitution prescribes the voting procedures in Section 8 “Voting,” Article VI “Voting System.” In practice, the room is set up with the House of Representatives, or State-designated representatives, on the left facing the front of the room, and the House of Delegates, or State and local officials, on the right. Behind them, at the back of the room, are the non-voting delegates and the associate and advisory members. A voting item is passed if a minimum of 27 members of the House of Representatives votes yea *and* if a majority of the members of the House of Delegates votes yea – a minimum of 27 yea votes are required.[13]

Currently, there is a combined membership of approximately 2,400.

The four regional associations represent all 50 states, the District of Columbia and two U.S. territories.[14]

The regional associations meet annually. They are organized the same as the NCWM with standing committees to consider changes to model laws, regulations and national standards published by NIST in a series of Handbooks. The NCWM has several technical standing committees: Laws and Regulations (L&R), Specifications and Tolerances (S&T) and Professional Development Committee (PDC). The NCWM meets twice each year at interim and annual meetings. Each standing committee is comprised of five members appointed by the Chairman of the NCWM for a five-year term. Members must represent all four regional associations.

All voting on amendments to the NIST Handbook series is done at the annual meeting. The proceedings of the NCWM are compiled in a NIST special publication series. For example, the 2019 meeting of the NCWM was documented in NIST Special Publication 1253. *Report of the 104th National Conference on Weights and Measures. Milwaukee, Wisconsin, July 14–18, 2019* (August 2020). Voting items are shown with proposed changes and both the state directors and inspectors must

Table 4.1 NCWM Membership (June 2019)

Membership	*Count*
State Government	805
Local Government	464
Total Government	1,269
NIST	15
Other Federal Government	10
Foreign Government	14
Retired	232
Total Advisory	271
Associate	768
Foreign Associate	96
Total Associate	864
Grand Total	2,404

Source: NIST Special Publication 1253. Report of the 104th National Conference on Weights and Measures. Milwaukee, Wisconsin, July 14–18, 2019. (August 2020), "BOD 2019 Final Report" pp. BOD-3 and 4.

Table 4.2 Regional Associations of the NCWM

Central Weights and Measures Association (CWMA)	*Northeastern Weights and Measures Association (NEWMA)*	*Southern Weights and Measures Association (SWMA)*	*Western Weights and Measures Association (WWMA)*
Illinois, Indiana, Iowa, Kansas, Michigan, Minnesota, Missouri, Nebraska, North Dakota, Ohio, South Dakota, Wisconsin	Connecticut, Maine, Massachusetts, New Hampshire, New Jersey, New York, Pennsylvania, Puerto Rico, Rhode Island, Vermont	Alabama, Arkansas, Delaware, District of Columbia, Florida, Georgia, Kentucky, Louisiana, Maryland, Mississippi, North Carolina, Oklahoma, South Carolina, Tennessee, Texas, U.S. Virgin Islands, Virginia, West Virginia	Alaska, Arizona, California, Colorado, Hawaii, Idaho, Montana, Nevada, New Mexico, Oregon, Utah, Washington, Wyoming

vote to pass an item. Generally, state directors are political appointees and they tend to be more receptive to changes sought by the regulated industry than the weights and measures inspectors. For example, since inspectors conduct the field inspections, they know first-hand the difficulty of achieving effective weights and measures regulation when standards are diluted by exemptions.

Model law or standard changes only become effective when a state adopts it into statute. Local government (counties and municipalities) adopts state weights and measures law into local ordinances to allow enforcement action by their inspectors.

The Goal of Uniformity

In a word, '*uniformity*' is the principal goal of the many state and local government weights and measures regulatory programs. This purpose is set out in the Introduction of NIST Handbook 130. *Uniform Laws and Regulations in the area of legal metrology and engine fuel quality* (2020):

> The purpose of these Uniform Laws and Regulations is to achieve, to the maximum extent possible, uniformity in weights and measures laws and regulations among the various states and local jurisdictions in order to facilitate trade between the states, permit fair competition among businesses, and provide uniform and sufficient protection to all consumers in commercial weights and measures practices.[15]

Uniform national weights and measures laws, technical standards and inspection procedures:

(1) serve to *level the playing field* for competing producers, and (2) allow *value comparisons* by consumers and make certain that they receive accurate weights and measures. However, NIST, NCWM and state W&M programs have had only limited success in achieving uniformity. A few examples here will probably suffice.

Uniformity of Weights and Measures Laws, Regulations and Standards

First, it will help to elaborate on what is meant by *uniformity*. In the broadest terms, it means that the 50 states will adopt common weights and measures laws, regulations and standards; develop standardized inspection practices for devices and packages; and provide for uniform inspector training and certification. State adoption of common weights and measures law, regulations and standards are fairly simple to assess since NIST publishes an up-to-date table, "Uniformity of Laws and Regulations," at the front of each annual edition of Handbook 130 *Uniform Laws and Regulations: in the areas of legal metrology and engine fuel quality.* The adoption, by states and territories, of model weights and measures laws and regulations in 2012 and 2020 are compared in Tables 4.3 and 4.4.

The sum of 'YES' and 'yes' indicate how many states and territories have adopted a current or previous NCWM model law, regulation or standard. Overall, there has been little progress from 2012 to 2020, perhaps even some regression, except for a tripling of state adoption of price verification regulation (it was still pretty new in 2012). There are 53 states and territories but, as of 2020, only 8 out of 13 NCWM model laws, regulations or standards have been adopted by even a *majority* of states and territories. By this measure, the achievement of uniformity of laws, regulations or standards is not progressing. See Table 4.5.

The National Conference on Weights and Measures (NCWM) has acknowledged the problems caused by the lack of uniform weights and

Table 4.3 Uniformity of Laws and Regulations: 50 States, D.C., Puerto Rico and Virgin Is. [2012]

Laws, Regulations and Standards	*YES*	*yes*	*yes**	*NO*	*no*
Weights and Measures Law	4	43	5	0	1
Weighmaster Law or Regulation	0	21	10	21	1
Uniform Fuel Law	2	8	35	8	0
Packaging and Labeling Regulation	19	29	4	1	0
Method of Sale Regulation	18	27	5	2	1
Price Verification Regulation	8	2	3	9	2
Unit Pricing Regulation	5	5	11	29	3
Registration of Service Agents Regulation	4	29	14	7	0
Open Dating Regulation	5	3	9	32	3
Type Evaluation Regulation	12	28	4	4	5
Uniform Engine Fuel Regulation	2	9	32	8	2
Handbook 44 Standards	41	11	1	0	0
Handbook 133 Standards	32	12	0	2	7

Notes

YES – Adopted and updated on an annual basis
Yes – Law or regulation from earlier year
Yes* - Law or regulation but not based on NCWM
NO – No law or regulation
No – No law/regulation but NCWM used as guideline

Source: NIST Special Publication 1253. *Report of the 104th National Conference on Weights and Measures. Milwaukee, Wisconsin, July 14–18, 2019.* (August 2020), "BOD 2019 Final Report" p. 13.

Table 4.4 Uniformity of Laws and Regulations: 50 States, D.C., Puerto Rico and Virgin Is. [2020]

Laws, Regulations and Standards	*YES*	*yes*	*yes**	*NO*	*no*
Weights and Measures Law	4	40	8	0	1
Weighmaster Law or Regulation	0	21	10	21	1
Uniform Fuel Law	3	7	37	6	0
Packaging and Labeling Regulation	19	29	4	1	0
Method of Sale Regulation	18	27	6	2	0
Price Verification Regulation	18	16	10	8	1
Unit Pricing Regulation	5	4	12	28	4
Registration of Service Agents Regulation	5	25	15	8	0
Open Dating Regulation	5	3	9	33	3
Type Evaluation Regulation	11	29	6	3	4
Uniform Fuel Regulation	5	11	30	5	2
Handbook 44 Standards	36	16	1	0	0
Handbook 133 Standards	26	19	0	2	6

Notes

YES – Adopted and updated on an annual basis
Yes – Law or regulation from earlier year
Yes* – Law or regulation but not based on NCWM
NO – No law or regulation
No – No law/regulation but NCWM used as guideline

Source: NIST Special Publication 1253. *Report of the 104th National Conference on Weights and Measures. Milwaukee, Wisconsin, July 14–18, 2019.* (August 2020), "BOD 2019 Final Report" p. 9.

measures laws and regulations among the states in their publication, NIST Handbook 155 *Weights and Measures Program Requirements: A Handbook for the Weights and Measures Administrator*.

> The decentralized weights and measures system in the United States creates a great challenge to achieve uniformity among the many regulatory jurisdictions. Conflicting regulations, varying interpretations of the same or similar requirements and divergent methods of enforcement seriously interfere with the efficiency of any program and are particularly unfortunate when associated with the administration of a weights and measures program. In the first place such conflicts and disparities throw a great burden upon the manufacturers of weighing and measuring equipment, a burden that is eventually borne by the ultimate consumer through increased costs of the product he buys. In the second place these are most confusing to the business interests of the state, which are forced to conform to whatever requirements may be in force in the

Table 4.5 Uniformity of Laws and Regulations: Change 2012–2020

Laws, Regulations and Standards	*YES+yes 2012*	*Yes+yes 2020 (Percent)*	*2012–2020 Change (Percent)*
Weights and Measures Law	47	44 (83.0%)	-3 (-6.4%)
Weighmaster Law or Regulation	21	21 (39.6%)	0
Uniform Fuel Law	10	10 (18.9%)	0
Packaging and Labeling Regulation	48	48 (90.6%)	0
Method of Sale Regulation	45	45 (84.9%)	0
Price Verification Regulation	10	34 (64.2%)	+24 (2+40.0%)
Unit Pricing Regulation	10	9 (17.0%)	-1 (-10.0%)
Registration of Service Agents Regulation	33	30 (56.6%)	-3 (-9.1%)
Open Dating Regulation	8	8 (15.1%)	0
Type Evaluation Regulation	40	40 (75.5%)	0
Uniform Engine Fuel Regulation	11	16 (30.2%)	+5 (+45.5%)
Handbook 44 Standards	52	52 (98.1%)	0
Handbook 133 Standards	44	45 (84.9%)	+1 (2.3%)

Notes

YES – Adopted and updated on an annual basis
Yes – Law or regulation from earlier year
Yes* – Law or regulation but not based on NCWM
NO – No law or regulation
No – No law/regulation but NCWM used as guideline

locality where a particular transaction takes place. Also, non-uniform requirements confuse the purchasing public, and complicate the enforcement of the law and hamper the officials who are trying to enforce it.[16]

Uniformity of Weights and Measures Device Categories

The state surveys of weights and measures programs conducted in 2002 and 2019, examined thoroughly in Chapter 1, are stark evidence of the lack of uniformity, among states, in defining device categories. The confusion of different device categories among states makes it nearly impossible to even count the number of weighing and measuring devices in the United States. This frustrates comparisons, between states, of device inventories, inspections, failure rates and other metrics of state weights and measures regulatory programs. This means that there can be no meaningful feedback loop to allow the NCWM to assess whether existing model laws, regulations or device codes are effective or whether amendments are needed.

The lack of uniformity in weighing and measuring device categories undermines virtually all other efforts to standardize weights and measures requirements among the states. As a result, this unnecessarily complicates business practices by regional and national companies that operate in more than one state.

Uniformity of Weighing and Measuring Device Inspection Forms

The NCWM has promoted the use of several standardized weighing and measuring device inspection forms but states, by and large, prefer their own. This impedes efforts to standardize databases, streamline data entry, or sort device inventory and inspection records in databases so that it can be easily compared to previous years or programs in other states.

Uniformity of Inspection Procedures

Inspection procedures differ considerably among states. For example, with respect to retail motor-fuel dispenser inspections, states often have very different policies regarding: station average (errors on over-registration), slow fill tests, sealing meters, fuel quality tests, electronic audit trails, checking for credit card skimmers, recording data (e.g., totalizer readings, inspection 'time in' and 'time out') and the role of service companies. Inspection procedures for package net contents and price scanning systems – if the states even conduct these inspection activities – have been discussed previously and are in similar disarray.

Uniformity of Training and Certification of Weights and Measures Inspectors

The NCWM has been developing a new inspector certification program but it has limitations and the NCWM is still standing up the Professional Certification Exam program. Since 2010, a growing list of professional certification exams and basic competency exams for inspectors have become available online. The Professional Development Committee (PDC) of the National Conference on Weights and Measures (NCWM) reported, at the 104th National Conference on Weights and Measures in Milwaukee, Wisconsin during July 14–18, 2019, that 1,042 exams had been passed between fiscal year 2010–2011 and fiscal year 2017–2018.[17]

Presently, there are nine professional certification exams. The most recent additions include: LPG and Anhydrous Ammonia Liquid Meters, Precision Scales and Price Verification. There are also two basic

Table 4.6 Professional Certification Exam Program - Training Certificates Issued by NCWM

Professional Certification Exam	*Certificates Issued* [a]
Retail Motor-Fuel Dispensers (2010)	364
Small Capacity Scales (2012)	249
Package Checking – Basic (2012)	167
Medium Capacity Scales (2015)	112
Large Capacity Scales (2015)	74
Vehicle-Tank Meters (2015)	76

Note

[a] Totals may include expired certificates. Professional Certificates must be renewed every five years.

competency exams (prerequisites): Basic Weighing Devices and Basic Liquid-Measuring Devices. All exams are proctored. The exam fee for non-members is $75. Examinees are permitted to use applicable NIST Handbooks and Examination Procedure Outlines (NIST Handbook 112) during the exams.

The Professional Certification Exam for Retail Motor-Fuel Dispensing Systems consists of three timed sections (a total of 2 hours) and 50 questions (primarily multiple choice). The passing score is 70 percent. Publications allowed are NIST Handbook 44 *Specifications, Tolerances, and Other Technical Requirements for Weighing and Measuring Devices* (Section 3.30 "Liquid-Measuring Devices") and NIST Handbook 112 *Examination Procedure Outlines* (EPO 21 – "Retail Motor-fuel Dispensers," EPO 22 – "Retail Motor-Fuel Dispensers Blended").

NIST training resources are limited to a few train-the-trainer courses each year and training webinars online. Clearly, this is inadequate to meet the needs of state and local weights and measures regulatory programs. The Professional Development Committee, a standing committee of the National Conference on Weights and Measures, explains the paucity of training sponsored by NIST as follows:

> The [PDC] Committee has reiterated multiple times in the past that the responsibility for training employees rests with individual organizations (weights and measures jurisdictions and industry alike) …The Committee recognized that NIST OWM [Office of Weights and Measures] cannot possibly train all weights and measures inspectors in the country. *The state and municipal jurisdictions have ultimate responsibility for training and qualifying their personnel.*[18] [*italics* added]

The statement by the PDC that, "*The state and municipal jurisdictions have ultimate responsibility for training and qualifying their personnel,*" is the principal reason for the lack of uniformity in the training and certification of weights and measures inspectors. NIST/OWM may not have the resources (staff, budget) to train state and local government weights and measures inspectors but state and local government weights and measure programs do not either. State and local government weights and measures program budgets are very limited. Growing inspection workloads and shrinking inspector staff mean that there is not sufficient time available for either classroom training or follow-up on-the-job training. Virtually no new-hire inspectors come to the job with experience in weights and measures so all inspectors must be trained and certified before they can conduct inspections without supervision. It can take a year or more to become qualified in a few basic inspection types such as retail motor-fuel dispensers (gas pumps), small capacity scales and price scanning systems. Inspectors are generally assigned a territory and they are responsible for all device and package inspections therein. This means that considerable training is required before these inspectors have the training that they need to assume full responsibility for a territory. Lastly, few weights and measures jurisdictions have inspectors on staff that have the training skills, knowledge and experience that they need to be effective trainers.

Memorandum of Understanding

The National Conference on Weights and Measures (NCWM) and National Institute of Standards and Technology (NIST) have entered into a Memorandum of Understanding (MOU) on their working relationship. The MOU, signed in 1998, reads, in part, as follows:

> This Memorandum of Understanding reaffirms the cooperation between the National Conference on Weights and Measures, Inc. (NCWM, Inc.) and the National Institute of Standards and Technology (NIST) "in securing uniformity in weights and measures laws and methods of inspection" in accordance with the National Institute of Standards and Technology Act. Starting in 1905, the Parties and their legal predecessors have jointly conducted annual meetings, which have come to be known as the National Conference on Weights and Measures (the "Conference"). The Parties agree to preserve this historical relationship through their continued joint sponsorship of the Conference. Traditionally, the Director of NIST has served as the Honorary President of the National Conference

on Weights and Measures, and the Chief of the NIST Office of Weights and Measures has served as the Executive Secretary of the Conference.[19]

Notes

1 Rexmond C. Cochrane. *Measures for Progress: A History of the National Bureau of Standards*. U. S. Department of Commerce, National Bureau of Standards. (1966), Appendix C "Basic Legislation: Relating to Standards of Weights and Measures and to the Organization, Functions, and Activities of the National Bureau of Standards", p. 541.

2 Department of Commerce and Labor, Bureau of Standards. *First Conference on Weights and Measures of the United States*. Letter of Submittal for the "Proceedings of the First Annual Meeting of the Sealers of Weights and Measures of the United States: Held at the Bureau of Standards, Washington, D.C. January 16 and 17, 1905" [Reprint by Forgotten Books Ltd., London (2018)], p. 3.

3 Op. Cit., Cochrane, p. 87.

4 Department of Commerce and Labor, Bureau of Standards. *Second Annual Conference on the Weights and Measures of the United States*. Letter of Submittal for the "Report of the Second Conference on Weights and Measures: Held at the Bureau of Standards, Washington, D.C., April 12 and 13, 1906" [Reprint by Forgotten Books Ltd., London (2018)], p.3.

5 Op. Cit., Cochrane, pp. 87–88.

6 Department of Commerce and Labor, Bureau of Standards. *Third Annual Conference on the Weights and Measures of the United States*. Report of the Third Conference on Weights and Measures: Held at the Bureau of Standards, Washington, D.C., May 16 and 17, 1907" [Reprint by Forgotten Books Ltd., London (2018)], pp. 107–110.

7 Department of Commerce, Bureau of Standards. *Tolerances and Specifications for Weights and Measures and Weighing and Measuring Devices. As adopted by the Tenth Annual Conference on the Weights and Measures of the United States Held at the Bureau of Standards, Washington, D.C. May 25–28, 1915*. Washington, D.C.: Government Printing Office (1915). This publication was reprinted as a commemorative edition to celebrate the 100th anniversary of the National Conference on Weights and Measures (1905–2005). 29 pages.

8 Department of Commerce and Labor, Bureau of Standards. *Weights and Measures: Fourth Annual Conference of Representatives from Various States*. Report of the Fourth Conference on Weights and Measures: Held at the Bureau of Standards, Washington, D.C., December 17 and 18, 1908" [Reprint by Forgotten Books Ltd., London (2018)], pp. 57–59, 73–75.

9 Op. Cit., Cochrane, pp. 88–89.

10 Ibid., p.90.

11 National Institute of Standards and Technology. *Handbook 155. Weights and Measures Program Requirements: A Handbook for the Weights and Measures Administrator*. (2011), p. 25.
12 National Institute of Standard and Technology. Special Publication 845. *Report of the 77th National Conference on Weights and Measures*. Nashville, Tennessee (1992) Report of the Executive Committee, Appendix E, p. 122.
13 National Institute of Standards and Technology. Special Publication 854. *Report of the 78th National Conference on Weights and Measures*. Kansas City, Missouri (1993), Report of the Executive Committee, Appendix D, p. 111.
14 Ibid., p. xxii.
15 National Institute of Standards and Technology (NIST). Handbook 130. *Uniform Laws and Regulations in the area of legal metrology and engine fuel quality* (2020), p. 1.
16 NIST Handbook 155. Op. cit., p.11.
17 National Institute of Standards and Technology (NIST). Special Publication 1253. *Report of the 104th National Conference on Weights and Measures* (August 2020), "PDC 2019 Final Report", p. PDC-4.
18 Ibid., p. PDC-13.
19 Memorandum of Understanding Between the National Conference on Weights and Measures, Inc. and the National Institute of Standards and Technology (July 14, 1998).

Bibliography

Cochrane, Rexmond C. *Measures for Progress: A History of the National Bureau of Standards*. Washington, D.C.: U. S. Department of Commerce, National Bureau of Standards. (1966).

Department of Commerce, Bureau of Standards. *Tolerances and Specifications for Weights and Measures and Weighing and Measuring Devices. As adopted by the Tenth Annual Conference on the Weights and Measures of the United States Held at the Bureau of Standards, Washington, D.C. May 25–28, 1915*. Washington, D.C.: Government Printing Office (1915).

Department of Commerce and Labor, Bureau of Standards. *First Conference on Weights and Measures of the United States*. Letter of Submittal for the "Proceedings of the First Annual Meeting of the Sealers of Weights and Measures of the United States: Held at the Bureau of Standards, Washington, D.C. January 16 and 17, 1905" [Reprint by Forgotten Books Ltd., London (2018)].

Department of Commerce and Labor, Bureau of Standards. *Second Annual Conference on the Weights and Measures.* Held at the Bureau of Standards, Washington, D.C., April 12 and 13, 1906 [Reprint by Forgotten Books Ltd., London (2018)].

Department of Commerce and Labor, Bureau of Standards. *Third Annual Conference on the Weights and Measures of the United States*. Report of the Third Conference on Weights and Measures: Held at the Bureau of Standards, Washington, D.C., May 16 and 17, 1907 [Reprint by Forgotten Books Ltd., London (2018)].

Department of Commerce and Labor, Bureau of Standards. *Weights and Measures: Fourth Annual Conference of Representatives from Various States*. Report of the Fourth Conference on Weights and Measures: Held at the Bureau of Standards, Washington, D.C., December 17 and 18, 1908 [Reprint by Forgotten Books Ltd., London (2018)]

National Institute of Standard and Technology. Special Publication 845. *Report of the 77th National Conference on Weights and Measures*. Nashville, Tennessee: National Institute of Standards and Technology (1992).

National Institute of Standards and Technology. Special Publication 854. *Report of the 78th National Conference on Weights and Measures*. Kansas City, Missouri: National Institute of Standards and Technology (1993).

National Institute of Standards and Technology. Handbook 155. *Weights and Measures Program Requirements: A Handbook for the Weights and Measures Administrator.* National Institute of Standards and Technology (2011).

National Institute of Standards and Technology. Special Publication 1253. *Report of the 104th National Conference on Weights and Measures.* Milwaukee, Wisconsin, July 14–18, 2019. (August 2020).

5 Enforcement Issues

This chapter presents several enforcement issues that have had a profound impact on weights and measures regulation in the United States. This is clearly not a comprehensive list. Rather, these examples are illustrative of the broad range of issues that challenge management of weights and measures regulatory programs today.

Joint Federal-State Study of School Milk (April–May 1997)

On July 17, 1997, the Federal Trade Commission (FTC) issued a report, "Milk: Does It Measure Up?," on the joint federal-state survey of the accuracy of net contents declarations on the labels of dairy products at schools, retail stores, state/federal facilities (hospitals, universities) and dairies. The report was prepared by staff from the Federal Trade Commission (FTC), U.S. Department of Agriculture (USDA), U.S. Food and Drug Administration (FDA) and the National Institute of Standards and Technology (NIST). The package inspections cited in the survey were conducted by weights and measures inspectors from 20 states: Alabama, California, Delaware, Florida, Iowa, Kansas, Louisiana, Maryland, Massachusetts, Minnesota, Mississippi, Montana, New York, Oklahoma, Tennessee, Texas, Utah, Washington, West Virginia and Wisconsin. The inspections were carried out in a three-week period during April–May 1997.

Study Findings

The report concluded that the survey found, "widespread problems with short-filling of milk, other dairy products [yogurt, cottage cheese] and juice."

Just over 40 percent of the 1638 inspection lots failed. Of the 858 lots of milk and juice inspected at schools, universities and hospitals, almost

DOI: 10.4324/9781003263661-6

Table 5.1 School Milk Survey (1997): Inspection Lot Results

Location Type	*Sites Visited*	*Passed Lots*	*Failed Lots*
Schools	264	388 (52%)	364 (48%)
State/Federal Facilities	32	59 ((56%)	47 (44%)
Retailers	138	298 (68%)	142 (32%)
Dairies/Packagers	78	227 (68%)	113 (33%)
Total	512	972 (59%)	666 (41%)

Source: Federal Trade Commission. "Milk: Does It Measure Up?: A Report on the Accuracy of Net Contents Labeling of Milk and Other Products" (July 17, 1997).

one-half failed inspection. Of the 780 lots of milk and dairy products inspected in retail stores, packaging plants and dairies, almost one-third failed inspection.[1]

The results of the package inspections are summarized in Table 5-1.[2] The percentage of inspection lots that failed to meet the net contents declarations on product labels was exceedingly high.

A total of 1,292 lots of milk were inspected: 701 lots (54%) passed and 591 lots (46%) failed. More than one-half of all lots of school milk in 8 oz. (half-pint) containers failed to meet net contents declarations: 356 lots (48%) passed and 391 lots (52%) failed.[3]

Impact on Consumer

Given these short-filled package statistics, the overcharge to consumers of milk products was estimated to be very significant.

> In 1997, NIST staff assisted a collaboration of 40 states, the USDA, the FDA, and the FTC in investigating school prepackaged milk. They found that 45% of the containers were short filled, at a cost to consumers of nearly $30 million.[4]

According to the U.S. Department of Agriculture (USDA), milk production in the United States was 218.382 billion pounds in 2019.[5] One gallon of whole milk is assumed by the USDA Economic Research Service to weigh 8.6 pounds. Actually, the weight of a gallon of milk varies depending on its density. Whole milk is heavier than skim milk. The average price of a gallon of whole milk in 2019 was $3.32.[6] Clearly, the dairy industry is a significant part of the economy in the United States. The industry is active in all 50 states.

Even a small short-measure error per unit is multiplied. The errors found during the 1997 milk survey ranged from "less than 1% to 6%

or more."[7] There is an old saying among veteran weights and measures inspectors that is still apt, "Weights and measures saves millions of dollars for consumers – one nickel at a time." More than 40 percent of 1,638 inspection lots in the milk survey failed.

> When inspection lots of milk and other products were rejected, the average amount of shortage per package ranged from less than 1% to 6% or more. Although these shortages represent only a small amount per individual package of milk or juice, the aggregate shortages represent a substantial amount of product over time. This causes economic losses to consumers and major purchasers, such as school districts, hospitals and universities. Further, this short-filling affects the milk and juice served with school breakfasts and lunches. In addition, retailers, wholesalers and dairies experience business disruptions and sales losses when short-filled products are removed from sale by government inspectors. Furthermore, injury to competition may result from inequities in the marketplace caused by short-filling of packages by some industry members.[8]

Milk Net Contents Inspection Procedures

The author participated in the 1997 milk survey as the manager of the weights and measures regulatory program in Seattle, Washington.[9] The author managed an active package inspection program including production sites (e.g., dairies, soft drink bottling plants, coffee breweries, bakeries, pre-mix drywall joint compound, mulch), wholesale warehouses and retail stores (e.g., supermarkets, big box stores).

A typical package inspection at a local dairy required about two or three hours for two inspectors. After meeting with the dairy manager, inspectors moved to the cold room near the loading dock to specify an inspection lot. For example, the inspectors may select half-gallon packages of 2 percent chocolate milk produced that day. The population of these packages in the cold room is counted to determine an appropriate sample size – typically a sample size of 12 or 24 packages. Sample packages are selected by using a random number table so that random sampling procedures are applied – similar to procedures for random sampling in opinion polls. Two or three packages are selected from the sample for destructive testing (i.e., package is opened and cannot be sold). A 'gravimetric' testing procedure was used to convert the volume of one half-gallon of milk to a weight. The milk is poured into a half-gallon volumetric test measure from the inspector's set of calibrated glassware. The milk from a second package is used, if necessary, until the milk level rests on the glassware half-gallon mark. The temperature of the milk is verified to be within the permitted range (typical refrigerator temperature), with a calibrated thermometer, because the glassware test measure expands or contracts at different temperatures. Previous to filling the glassware volumetric test measure with milk, the glassware is filled and emptied to wet the internal surface since there is always a small amount of *clingage* remaining when it is emptied. Then the glassware weighed on the inspector's calibrated portable scale and zeroed so that the test measure is deducted like tare. As a result, when the filled test measure is weighed, the weight displayed on the scale is the net weight of one-half gallon of milk. The '*target weight*' is the net weight plus the average unused dry tare weight (package and screw cap). Next the target weight is zeroed so it is treated as tare. Then, the remaining packages in the sample are weighed. The weights of the remaining sample packages are filling errors – either plus (overfill) or minus (short measure). A worksheet[10] is completed by the inspector to compute the average package error. The average error for the sample must be positive or zero or the entire inspection lot fails and is ordered 'off sale.' The inspector applies a sample correction factor following the procedure outlined in NIST Handbook 133, *Checking the Net Contents of Packaged Goods*. These correction factors make allowances for the small size of the sample so that it accurately represents the entire population. The packages in a failed lot cannot be sold until relabeled with the actual net weights. For example, the milk packages in the failed lot could be used to make ice cream or they could be donated. In that case, the inspector witnesses the empty packages, or the manifest donating the milk, in order to verify that the milk packages are not sold in the marketplace.

If a milk lot fails inspection, there are normally a few common causes. Most often, the dairy workers operating the filling equipment failed to use the 'check weigher' (in line scale) to verify that the first few packages through the filling equipment either met or exceeded the target weight for that milk product. Another cause of short measure could be that the filling equipment is operated too fast. As a result, *foaming* from filling the containers too fast causes the equipment to erroneously indicate that the containers are filled before they are. Generally, the *average* package in the sample must meet or exceed the declared net contents on the label of the milk container. However, a lot will also fail if there are individual packages that have unreasonable errors called the 'maximum allowable variation' (MAV). MAVs indicate that the packaging process is not sufficiently reliable.

NIST Handbook 133 lists the "maximum allowable variation", or MAV, for different labeled contents. For example, for half-pints, the MAV is 0.38 fluid ounces, and for half-gallons, the MAV is 1.5 fluid ounces. For lots consisting of 3,200 packages or less, if a single package in the random sample exceeds the MAV, the lot fails inspection. For lots of more than 3,200 items, the handbook permits one package in the random sample to exceed the MAV.[11]

Political Fallout from the School Milk Study

What is perhaps most noteworthy about the 1997 milk survey is how it strained the traditionally close working relationship between the National Conference on Weights and Measures (NCWM) and the federal National Institute of Standards and Technology (NIST). In most states, the weights and measures program is part of the state department of agriculture. State departments of agriculture have, as one of their principal functions, the promotion of the state agricultural sector including the dairy industry. The milk survey gave the dairy industry, in the 20 participating states, a *black eye* because of the widespread short measure found. The resulting negative media coverage embarrassed political appointees leading the state departments of agriculture and they tended to hold both the state weights and measures programs and NIST responsible – instead of the industry that short-filled the school milk containers! The fallout due to the negative media coverage of the milk survey led directors of the state weights and measures programs, in concert with regulated industry, to propose a new "Protocol for Conduct of National Studies" (NCWM Board of Directors Item 101–14) at the annual National Conference on Weights and Measures convened in 1999.[12] The protocol was adopted and, notably, it required NCWM approval for future marketplace studies. The protocol, in some cases, even provided for advance notification of the affected industries. This was a profound change. Notification was contrary to procedures established for all weights and measures compliance inspections which are unannounced for obvious reasons. Based on these restrictions inserted in the marketplace study protocol, it appears possible that one purpose may have been to discourage the conduct of any national surveys in the future. This is especially likely since the protocol was proposed in the immediate aftermath of the school milk study.

Marketplace Surveys After the School Milk Study

As if to confirm that the intent of the new marketplace study protocol was to discourage future surveys, only a few marketplace surveys have been conducted since the new protocol was put in place. There was a survey of store-packed random-meat and poultry items in 14 weights and measures jurisdictions coordinated by staff from the California Division of Measurement Standards during 2006.[13] There was also a multi-state seafood investigation, coordinated by the Chief of the Wisconsin Department of Agriculture, Trade and Consumer Protection, with participation by 17 states during January 18–February 12, 2010.[14] The NCWM could not manage the latter since it was a regulatory investigation (law enforcement) and not a so-called 'marketplace survey.' The Wisconsin investigation was intended to assess the scope of the problem where ice glaze on seafood is improperly included in the labeled net weight declaration. According to the investigation report, "Consumers purchase $22.7 billion of seafood to eat at home, and food services purchase $46.6 billion every year. If 2% of the weight of the $69.3 billion of seafood purchased annually is ice, annual loss would be $1.4 billion."[15] Packers routinely apply an ice glaze to seafood prior to packaging to protect the seafood during storage and distribution. The Wisconsin investigation found widespread overcharges due to the unlawful inclusion of ice glaze in determination of net weight of the seafood. In some packages examined, the ice glaze added $23 per pound to the price of a package of seafood. Inspectors ordered 21,000 packages of seafood off sale due to short weight during the investigation. Ice glaze added as much as 25 percent to the labeled weight of some packages of seafood.[16]

Implementation of Recommendations from Marketplace Studies

Marketplace surveys invariably demonstrate that more oversight of the packaging process is necessary by both production facilities and weights and measures inspectors. There are serious difficulties impeding weights and measure inspections to check the net contents of packaged goods. The number of states that conduct *any* package inspections has declined from 90.0 percent in 2002 to just 63.6 percent in 2019. This phenomena is primarily due to the 25.9 percent decline of inspector resources during that period. See Table 1.9. In addition, as state weights and measures programs become self-funded through device registration fees, general fund monies have declined as a percentage of annual budgets from 44.4 percent in 2002 to just 19.4 percent of states in 2019. Device registration fee revenue cannot be used to subsidize package inspections so there is very little funding for package inspections. See Table 1.10. Also, local municipal or county weights and measures programs do not have the funding to conduct regular package inspections at large production facilities that produce huge volumes of packaged goods for national or international markets. Until a reliable funding mechanism is found to support package inspections, no expansion of package inspections is likely.

The quick move by the National Conference on Weights and Measures to establish a strict protocol for national marketplace surveys immediately after the 1997 milk survey may indicate a related problem. Even though the protocol was developed by regulatory officials, the associate membership (industry representatives) of the NCWM numbers approximately two-thirds of the state and local government membership. The influence on the voting membership from regulated industry inside the NCWM has grown considerably over time. The NCWM is a standard-setting body comprised of weights and measures regulatory officials. It is debatable whether associate membership have an improper, albeit unintended, influence on NCWM decision-making regarding policies, model laws, or model regulations applicable to the regulated industry (the associate membership itself). For example, at a NCWM annual meeting in Burlington, Vermont, the general session discussion of a proposal by the plywood industry to allow the use of nominal (in name only) dimensions similar to nominal dimensions used by the dimensional lumber industry (e.g., 1.5"X3.5" actual, 2"X4" nominal). Many of the state directors supported the proposal while the local government weights and measures program managers and most weights and measures inspectors, were strongly opposed to the proposal as contrary to the basic principle that consumers must get what they pay for. The *exceptions* for specific products contained in Handbook 133 also seems contrary to the fair application of uniform weights and measures laws. For example, allowing premixed drywall joint compound to be sold by liquid measure instead of a solid. It is nearly impossible to verify net contents unless it is converted to a weight with conventional gravimetric procedures.

Retail Motor-Fuel Dispenser Fraud in Los Angeles (1998)

Overwhelmingly, errors in the quantity of a product delivered to a consumer in package form or from bulk, measured over a scale or through a meter, are *not* due to fraud. Instead, most errors result from equipment wear and human mistakes such as: (1) weighing and measuring devices are out of calibration, (2) package tare is inadequate, (3) shelf tags aren't updated to match the prices in the store computer or advertised prices, and (4) check weighers aren't used on production lines. Sometimes there is fraud. This is one such case. A big one.

Media Reporting

The NBC program *Dateline* aired a televised report, "Highway Robbery," on January 20, 1999, concerning a joint fuel dispenser fraud investigation by Los Angeles County Department of Agricultural Commissioner /Weights and Measures (ACWM) and the Los Angeles County District Attorney (LADA)/ Consumer Division.[17] The television report opened with an interview of a motorist who complained to Los Angeles Weights and Measures that his nearly empty vehicle fuel tank accepted 20.5 gallons even though the vehicle's owner manual stated that the fuel tank capacity was only 19.6 gallons. The motorist said that he had filled the tank until he could see the gasoline in the fill pipe. The motorist was concerned that the fuel dispenser was overstating the amount of gasoline that he actually received.

Historically, relatively few retail motor-fuel dispenser (gas pump) complaints are valid. Los Angeles County weights and measures receives approximately 1,000 gas pump complaints annually.[18] These complaints are investigated by uniformed weights and measures inspectors in county trucks with 5-gallon calibrated test measures that are used to test the accuracy of the delivery from a fuel dispenser. Vehicle owners are typically unaware of factors, such as overfilling, that affect the amount of fuel in a fuel tank.

> The vehicle's fuel tank capacity rating is a reasonable approximation…Some manufacturers estimate that fuel tank capacity can vary by as much as 3% from the actual tank capacity…
>
> The unusable volume is the portion of the fuel tank's liquid capacity that lies at the bottom of the tank out of reach of the fuel pump.
>
> The vehicle's fuel tank capacity does not include the vapor headspace (that portion of the tank compartment at a level above the filler pipe neck) or the volume of the filler pipe. Sometimes drivers ignore the pump nozzle's automatic shut-off and continue to hold the nozzle operating the lever open in an attempt to deliver additional fuel. When this happens, the additional fuel begins to fill the vapor headspace and the filler pipe that are not considered part of the tank's rated capacity, thus resulting in a delivery of fuel greater than the fuel tank capacity rating stated in the owner's manual. Similarly, the lanes that surround the service station pumps may not be level and fuel may shift into the vapor space thus allowing more fuel to be delivered into the tank.

> Another scenario that can prompt consumer complaints involves the fuel tank capacity and the fuel gauge indication. When the fuel tank indicates a particular level, the customer frequently assumes that this represents a corresponding fraction of the fuel tank capacity.
>
> A reserve amount of fuel is present in the tank if the manufacturer elects to set the fuel gauge to indicate "Full" at a level just below the tank's actual liquid capacity. Similarly, reserve fuel may be present in the tank if the manufacturer designs the fuel gauge to indicate "Empty" at a level above the actual point where the tank runs out of fuel.[19] [underline added for emphasis]

Subsequently, there were numerous fuel dispenser complaints at this and other gas stations. Weights and measures inspectors in Los Angeles County and adjacent counties conducted complaint investigations but, invariably, the fuel dispensers were found to be accurate.

Investigation Results

No formal written report was issued on the expanding investigation because it eventually became a Los Angeles District Attorney criminal investigation. However, in 1999, Kurt Floren, the current Agricultural Commissioner and Director of Weights and Measures for the County of Los Angeles, authored a four-page informal summary report, "Retail Petroleum Fraud Investigation." Kurt Floren was the lead investigator for ACWM for the fraud case which ultimately spanned almost four years (1994–1997). His report summarized the investigation as follows:

> Ultimately, we discovered 40 [gas] stations in 5 different counties run by 7 different operators, all of which had this technology installed. It was time to move…
>
> On a single day (October 29, 1997), 37 simultaneous search warrants were executed, involving approximately 60 weights and measures officials and 120 District Attorney investigators, the largest ever conducted by the L.A. DA Consumer Division office. Over 500 pulsers were seized from Gilbarco and Tokheim gas pumps, the manufacturers of which had also provided representatives to aid in the seizures and document observations.[20]

It is believed that this was the longest running and largest scope criminal investigation ever conducted by any weights and measures program in the United States. The weights and measures program in Los Angeles County today is one of the largest in the United States with 85 inspectors.

During the investigation, weights and measures inspectors employed two unmarked vehicles with concealed video equipment and concealed non-standard test measures. State of California weights and measures inspectors also conducted undercover buys with two unmarked vehicles similarly equipped. Los Angeles County Weights and Measures conducted an estimated 150 undercover buys during the day, late evening and on weekends.[21] Undercover 'planned buys' revealed that motor-fuel dispensers that had been tampered with yielded many shortages *greater than one gallon* on transactions of 7–9 gallons (13%) and 12–13 gallons (8%).[22]

> *To put things in perspective, Los Angeles County experiences approximately 4 billion gallons of retail fuel sales per year while, statewide, sales amount to over 15 billion gallons annually. Observed shortages in undercover purchases ranged from about 5% to as high as 30% depending upon the volume dispensed and the respective pulse addition via the fraudulent programming. The fraud potential was monumental.*[23]

The Original Short Measure Complaint (1994)

The retail motor-fuel dispenser (RMFD) fraud case actually started out as a 1994 short-measure complaint received by the California Division of Measurement Standards (DMS) that occurred in Riverside County. State inspectors conducted undercover vehicle purchases and found *significant* short-measure delivery. Short-measure errors found during routine, unannounced, annual compliance inspections are normally quite small – in the range of -6 to -8 cubic inches using a 5-gallon test measure with a sight glass. These are usually calibration errors due to normal wear of the retail motor-fuel dispenser (gas pump) as it ages in service. The tolerance allowed by NIST Handbook 44 is -6 cubic inches or about -0.5 percent. Large errors are very uncommon.

The same Riverside County gas station owner also operated a gas station in Los Angeles County. So, DMS requested that Los Angeles Agricultural Commissioner/Weights and Measures (ACWM) use its undercover cars to investigate that gas station as well. Both the two state (DMS) and two county (Los Angeles County ACWM) undercover vehicles were unmarked and equipped with concealed video equipment and concealed modified-volume test measures (not the 5-gallon standard test measure) to collect the evidence. Numerous instances of short-measure deliveries with unusually large errors were documented. As a result, in March 1995, joint teams of representatives from the Los Angeles County District Attorney (DA) and ACWM conducted simultaneous undercover purchases followed by the execution of simultaneous search warrants at two gas stations in Riverside County and one gas station in Los Angeles County. Subsequently, routine tests were conducted by uniformed weights and measures inspectors, using standard 5-gallon test measures, at the same gas stations but the tests did not reveal any short measure deliveries. Searches continued for eight hours but could not determine how the 'cheating' was accomplished or how it was turned off. The DA did not file charges against the operator of the gas stations for short-measure deliveries before the one-year statute of limitations expired because the ACWM requested that the investigation continue until they understood how widespread the fraud scheme was and how the 'cheating' was accomplished.[24]

Receipt of an FBI Tip (Early 1996)

Los Angeles County ACWM was passed a tip from the local FBI office that was purportedly received from an ex-employee of a gas station in LA County. The complainant claimed that the gas station was manipulating gas pumps to overcharge consumers. Subsequent investigations by ACWM, using undercover vehicles, confirmed that there were large-error short-measure deliveries. DMS conducted similar investigations at other gas stations operated by the same owner and found the same pattern of large-error short-measure deliveries.

Weights and measures inspectors searched ownership records to match gas stations with the same owners of the three gas stations known to be 'cheating' and then conducted investigations with undercover vehicles. They found a total of 12 gas stations operated by three different owners that fit the pattern of large-error short-measure deliveries. Weights and measures inspectors for DMS and ACWM believed that there must be a connection.[25]

The capacities of the fuel tanks concealed in the undercover vehicles, used to collect gasoline deliveries to use as evidence, were larger than the 5-gallon test measures used by uniformed weights and measures inspectors. This was to prove important. Investigations were carried out in two steps. First, an undercover vehicle purchased fuel at a gas station. If there was a significant short-measure delivery, a uniformed weights and measures inspector in a marked vehicle would test the same gas pump soon afterwards employing normal inspection procedures. The uniformed inspector informed the gas station owner that ACWM had received a consumer complaint about a short-measure delivery and the inspector was investigating it. The test draws by the inspector's 5-gallon test measures invariably proved to be within tolerances. DMS and ACWM inspectors believed that this indicated that the 'cheating' could be turned off. They still did not understand how the 'cheating' was done.

Finally, a Couple of Breaks (Late 1996–July 1997)

DMS advised that another gas station, this one in Kern County, operated by the same persons of interest, was found to have similar short-measure deliveries. A fuel meter in one gas dispenser had a service sticker from a service technician. After the ACWM conducted a laborious hand-search of paper inspection records and service company records, it was discovered that the same service technician had performed service at six other gas stations also under investigation for large-error short-measure deliveries.

About the same time, DMS received a complaint from a gas station operator in San Bernardino County alleging that something was wrong at a competing gas station which advertised suspiciously low prices. He complained that the other gas station, 'sold gas for the price that I pay to buy it.' DMS conducted undercover purchases and found the same large-error short-measure deliveries. A records check indicated that the service technician for the honest gas station owner was the same service technician for the gas station under investigation in Kern County. DMS and ACWM asked the honest gas station owner to request a meeting with the dishonest service technician and question him (on hidden video) about whether there was anything he could do to help him increase his profits. The dishonest service technician said that he could install a system to increase sales by 20 percent but it would cost about $10,000. In addition, the dishonest service technician said that he was to receive 15 percent of the fraud proceeds on an ongoing basis.[26] The amount of fraud was calculated as the difference between the readings on the console in the convenience store (includes gallons from adding pulses) and the totalizer reading on the gas pump (true measure of gallons pumped). The totalizer works much like the odometer in an automobile.

The DA investigators (sworn officers) obtained and provided $10,000 in cash and installed concealed video and listening devices in the office of the honest gas station owner to record the meeting with the dishonest service technician. The service technician was secretly recorded on video explaining how the fraud scheme worked. He said that he would

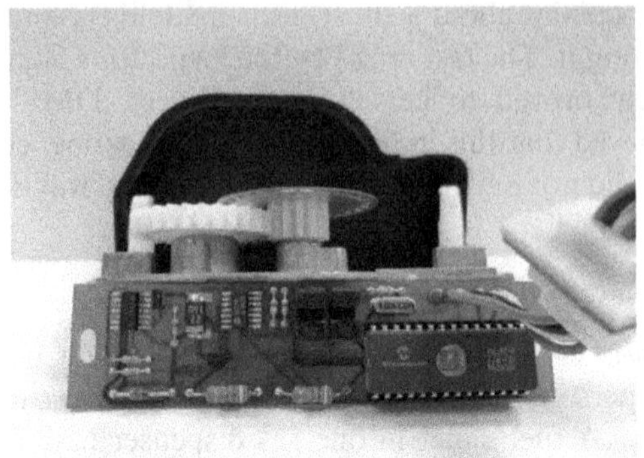

Photo 5.1 Gilbarco pulser unit with printed circuit board (PCB) removed. Microprocessor chip is reprogrammable.

Source: Photograph used by permission of County of Los Angeles, Department of Agricultural Commissioner/Weights and Measures.

install replacement printed circuit boards in the pulser units of the gas pumps containing microprocessor *chips* that had been reprogrammed to add *pulses* to indicate that more gasoline had been pumped than actually had. Modified pulser unit printed circuit boards were available for fuel dispensers manufactured by Gilbarco and Tokheim – major manufacturers of fuel dispensers. The pulser units in the gas pumps at the gas station operated by the honest owner were originally programmed with the ratio of approximately 1,000 pulses per gallon. The pulser unit converts mechanical motion of the gas pump meter to an electronic digital display on the gas pump, indicating total gallons pumped and total price.[27]

Setting the Trap (1997)

The service technician arrived at the gas station of the honest owner to install the reprogrammed computer chips on printed circuit boards (PCBs) replacing the original factory-installed PCB in the pulser units. DA investigators made video recordings of the process which required about three hours. The service technician explained to the honest owner that pulses, and therefore customer gallon indications and charge calculation, would be correct at 5 and 10 gallons because weights and measures inspectors use a 5-gallon test measure and took two draws (fast fill, slow fill). After the service technician finished and left, weights and measures inspectors tested the fuel dispensers – activating and

Photo 5.2 Gilbarco pulser unit assembled next to housing.

Note: The white plastic gear is driven directly by the fuel meter spindle. The gear drives the slotted wheel, which turns between infrared emitters (black blocks) on one side and pick-ups on the opposite side. Each time a spoke in the wheel breaks the infrared beam it is counted as a 'pulse.' The pulses are converted to 'gallons' then displayed on the fuel dispenser. Photograph used by permission of County of Los Angeles, Department of Agricultural Commissioner/ Weights and Measures.

Photo 5.3 Tokheim pulser unit PCB with microprocessor chip reprogrammed to 'cheat.'

Source: Photograph used by permission of County of Los Angeles, Department of Agricultural Commissioner/Weights and Measures.

deactivating the system by turning power to the fuel dispensers off and on with the emergency switch. 'Cheating' was either all 'on' or all 'off.'

ACWM and DMS weights and measures inspectors conducted many undercover purchases at suspect gas stations once that they understood how the fraud was accomplished. The concealed video in the under-cover vehicles documented an increase in the speed of advance of the 'gallons' display on the gas pumps when the 'cheating' started and a slow down when the 'cheating' stopped. The 'gallons' display sped up after 1.0 gallon, slowed after 4.0 gallons and sped up again after 5.0 gallons. The pattern repeated with the gallons display slowing down after 9.0 gallons and speeding up after 10.0 gallons.

The lead investigators from ACWM and DMS visited the manufacturers of the fuel dispensers installed at the gas stations under investigation (Gilbarco and Tokheim). They requested technical assistance in determining how the printed circuit boards in the pulser units had been modified to 'cheat.'

The Aftermath (1998–1999)

The Los Angeles County District Attorney charged seven persons with felony and misdemeanor crimes, civil violations and restitution. All of them ultimately pled guilty and no trials were conducted. The two leaders of the fraud scheme were charged with felony conspiracy and were sentenced to 360 days in confinement. One was a former 'insider' with Mepco Oil and the other was a computer programmer. They actually served one month in custody and the remainder in home detention. The service agent, who installed the modified printed circuit boards in the fuel dispensers that had been reprogrammed to 'cheat,' cooperated with prosecutors. He was also charged with felony conspiracy, served one day and was placed on three years of formal probation. Two gas station owner/operators were charged with misdemeanors, fined, placed on three years of formal probation and given 50 hours of community service. Two other gas station owner/operators were issued civil penalties. The president of Mepco Oil was required to pay $640,000 in penalties and costs.[28]

The Economic Impact of the Fraud Scheme on Consumers

It was estimated by Gil Garcetti, then Los Angeles District Attorney, that the criminals collected $1,000,000 while their fraud scheme was in place and that consumers were overcharged an average of $1 per transaction.[29]

> Dateline reported that cumulative overcharges to motorists from the fraud case could exceed $1,000,000 since it was estimated that average overcharges were $1.00 to $2.00 per transaction and there were one million or more transactions.[30]

In truth, the actual total losses are unknown. The average price per gallon of gasoline in Los Angeles, Orange, Riverside, San Bernardino and Ventura counties in 1997 was $1.34.[31] The average convenience store in the United States sold 142,500 gallons of gasoline products monthly in 2019.[32] If the average overcharge from the fraud was 10 percent, then a typical gas station could have collected fraudulent overcharges as follows: 100,000 gallons (rounded down for ease of computation) of gasoline monthly X 12 months per year X $1.34 per gallon X 0.10 = $160,800 annually. Since 40 gas stations were found to be 'cheating,' the total losses for the consumer due to overcharges could have been as much as $6,432,000 annually. If the 'cheating' started in 1994 and was stopped in 1997, then the losses would approach $25 million during that period. Unfortunately, it is not possible to learn the actual overcharges due to the fraud scheme because there were no reliable records. Mepco Oil was only fined $640,000 so it is possible that the fraud scheme was *very* profitable for the conspirators.

A review of the gas stations that were found to be 'cheating' revealed a pattern. Most of the gas stations were not major brand franchises. In addition, most gas stations were located in lower-income neighborhoods where transactions were primarily cash sales and not fill-ups. The maximum short measure fraud amounts occurred at about 4 gallons and 8 gallons, which roughly corresponded to $5 and $10 bills at the prevailing gas prices. No 'cheating' was done in small amounts, such as 1 gallon, since it would be too noticeable, e.g., filling motorcycles or gas cans for lawn mowers.[33]

A commentary that appeared in the *Loyola Consumer Law Review* in 2001 argued that the report, "Highway Robbery" produced by the television newsmagazine NBC *Dateline*, is inaccurate and misleading because it seeks to entertain rather than inform. As a result, it exaggerates and presents a one-sided view. For instance, the author cites an expert source in the National Institute of Standards and Technology (NIST) who was critical of the NBC *Dateline* piece for implying that there was little regulation. In response, the NIST representative is quoted as stating that, "there are thousands of weights and measures officials that go out every day and check dispensers for accuracy and performance." However, a nearly contemporary 2002 NCWM survey of the 50 state weights and measures regulatory programs only found a total of 1,835 inspectors who conduct annual inspections on 2.5 million retail motor-fuel dispensers, 1.0 million commercial scales[34] plus inspections on a broad array of other weights and measures devices (e.g., vehicle-tank meters, loading rack meters, liquefied petroleum gas meters, railroad track scales, belt conveyor scales, mass flow meters and taximeters) as well as checking package net contents, verifying accuracy of price scanning systems and many other activities. Inspection of a typical 24-pump gas station can require three hours and travel time between inspections is substantial outside of large cities. Inspectors are spread very thin. In 2002, there was only one inspector per 157,149 population or 1,928 square miles. See Table 1.2.

The author of the *Loyola Consumer Law Review* article also claimed that, "the computer systems used by gas stations today are difficult to tamper with..."[35] This is manifestly untrue in light of the Los Angeles fraud cases. In the Los Angeles fraud cases, the original pulser printed circuit boards were replaced with fraudulent PCBs without having to disturb the calibration mechanism, which bore a weights and measures wire-and-lead security seal affixed by the inspector. Modern gasoline dispensers have a key pad to change all parameters such as the 'mix ratio' where regular grade and premium are mixed to produce mid-grades. It is simple to produce a mid-grade that is 100 percent regular and no premium, instead of 50 percent of each, thereby collecting an extra 10 cents per gallon. The newest key pads are wireless and can change calibration remotely. The inspectors must rely on electronic audit trails to detect these types of fraud and these checks are rarely done due to inspector time and training constraints. The author of the article also reported information from some state and county weights and measures regulators stating that very few fuel dispensers ever failed inspection due to short measure in excess of tolerances.[36] The aforementioned 2002 NCWM survey of the 50 state weights and measures regulatory programs found that the fuel dispenser failure rate was 6.6 percent nationwide on 2.5 million retail motor-fuel dispensers. Lastly, the author claimed that cheating would be prevented since major oil companies require that "...gas station owners must maintain detailed records of all sales and deliveries, which are closely scrutinized by their franchisors, any discrepancy would be readily apparent."[37] In the Los Angeles fraud cases the gas station operators who were cheating understood this and kept two sets of books.

Implications for the Future

Today, nearly 25 years later, most modern weighing and measuring devices are electronic rather than mechanical. Many can be calibrated wirelessly. This bypasses wire-and-lead security seals affixed to the calibration mechanism of the device. A few examples are in order.

Tank trucks often make deliveries to gas stations two or three times per week to gravity-fill the underground storage tanks that supply gasoline products to the gas pumps. These trucks fill up at their own equivalent of a gas station where loading rack meters pump fuel from large above-ground storage tanks. When testing loading rack meters, the weights and measures inspectors usually pump 300 gallons to a portable test measure then read the error in the sight glass on the neck (fill pipe) of the test measure. Afterwards, if the meter is within tolerance, the inspectors affix a wire-and-lead security seal to the cover over the calibration mechanism. However, the control room for the busy facility has a computer system that can remotely and wirelessly change the calibration of the loading rack meters without disturbing the security seals. This voids the purpose of the security seals which is to prevent tampering to commit fraud. To compensate, the computer program has an electronic audit trail to show the initials of the person who changes any parameters of the loading rack meters, as well as the date and time. This offers some security but if passwords are compromised the effectiveness of this procedure is defeated.

The calibration mechanism for gas pumps at gas stations is located in a locked cabinet on the lower portion of the gas pump. The calibration mechanism is sealed by weights and measures inspectors, but service company agents often remove the security seals when conducting recalibration during routine periodic service calls. Duplicate keys to the cabinet locks are, in many cases, in the possession of service companies and weights and measures inspectors. In the upper part of the gas pump, a compartment is routinely opened to replace rolls of paper for printing receipts. There is a key pad inside that is used to change any parameters of the gas pump – e.g., selection of unit of measure (liters, gallons), changing the blend ratio (e.g., for a mid-grade product: usually 50% premium-50% regular). These parameters may also be changed remotely and wirelessly from the console by the clerk inside the convenience store.

The *Dateline* report, previously cited, stated that there were only 12 undercover vehicles in service in 1999 among 800 state, county and city weights and measures jurisdictions. They concluded that this will frustrate investigations like the one in Los Angeles where high tech was employed.[38] That seems to be a reasonable conclusion.

Weights and measures inspectors are always playing catch-up with new forms of potential fraud as technology improves since inspectors are not specialists in electronics or computer programming. Moreover, national test standards in the NIST handbooks take years to be developed and to be voted on at the regional and national conferences of the NCWM. For these reasons, it appears that enforcement will probably lag far behind technological innovations for the foreseeable future.

Privatization of Weights and Measures Enforcement

Budgets (adjusted for inflation) and the number of inspectors have both been declining for the past two decades according to trends identified by the 2002 and 2019 state weights and measures regulatory program surveys. The number of devices has increased during the same period. This indicates a need to prioritize or target inspections based on compliance histories or some other metric. However, many state weights and measures regulatory programs have no flexibility to appropriately shift their enforcement emphasis because of laws specifying annual inspections. In 80 percent of states, budgets have transitioned from the general fund to device registration fees and device inspection fees. Ironically, businesses expect their devices to be inspected annually *because* they pay these fees. Moreover, package inspections such as net contents and price scanning, have declined since, without General Fund monies, they have no funding source.

> Over the years, many jurisdictions have not been able to perform device inspections on an annual basis. In many cases, the reality is that the test cycle has been extended in spite of any time requirement specified in the law, because sufficient resources are not available to inspect all commercial measuring instruments on an annual basis. In some cases, weights and measures directors have effectively used the mandated test frequency to obtain additional funds to increase staff or obtain equipment to increase the efficiency of inspections to complete the annual testing of all devices.
>
> In other cases, when the weights and measures program brought the issue to state legislators that they were unable to perform inspections on all devices within the time frame specified in the law, the legislature changed the law to extend the inspection period for official inspections. The high compliance rates for accuracy and specification requirements for [small scales and] retail motor fuel dispensers (often greater than 95% and 90% respectively) often cause legislators to believe that cutbacks in the frequency of inspections can be made without significant consequences in compliance.[39]

Unfortunately, it is not understood by elected officials that failure rates are small precisely *due to* frequent compliance inspections. Since frequent inspections over many years have achieved high compliance rates, it often takes a few years before noncompliance rates inevitably begin to increase again. By then, the harm is widespread and slow to remedy.

Expanding the Role of Private Service Companies

One method pursued to try to maintain high compliance rates in the face of declining budgets and reduced inspector staff involves targeting devices with poor compliance histories for more frequent inspections. Jurisdictions must maintain databases of devices with compliance histories for this to work. However, some states don't have inventories of devices because they don't assess device registration fees. Another method involves inspection of samples of devices at one location but expanding inspections to all devices if the samples fail. Finally, another alternative is *privatization* of device inspections. In some states, this is limited to permitting registered service companies to recalibrate devices and place them back into service without awaiting reinspection by state inspectors.

> Any shift of what is ordinarily provided by government to the private sector is a form of privatization. There are many degrees of privatization. For example, many years ago, it was required that State or local government officials test all commercial devices after repair or installation. It is now routine for most States to permit registered repair firms to place new or repaired devices into service without immediate government testing. Registered service agencies should be perceived as an extension of weights and measures with the same responsibilities to apply the laws, test methods, and so on. This should be emphasized as a form of privatization that has already taken place across most of the nation.[40]

A few states have delegated annual or periodic inspections to service companies in part (e.g., commercial scales – Kansas) or entirely (e.g., New Hampshire) while conducting audits and investigating consumer complaints with a small number of state weights and measures inspectors. The term '*privatization*' means, in effect, surrendering government regulatory responsibilities to the private sector. This action, by state legislatures, is not without problems.

> One major criticism of this approach is that many view the service companies as having a conflict of interest when given a 'regulatory' responsibility to inspect and test measuring instruments and then they have the private sector responsibility to service the instruments and put them back into commercial service. Another criticism is that the government is abdicating its responsibility to oversee and regulate the commercial measurement system. To make

> this approach work, some states have required the owners of measuring instruments to have them tested annually.[41]

It is very important that private service company agents, permitted to conduct inspections in lieu of state inspectors, be subject to rigorous oversight. State inspectors must conduct follow-up inspections to audit the work of service companies. Service company agents must be thoroughly trained and properly licensed. The agents must be familiar with both state weights and measures law and NCWM standards for inspecting devices. Their field test equipment must be tested annually, and recalibrated as needed, by the state metrology lab to verify that it is accurate. Finally, service companies must submit device inspection records to the state for oversight purposes.

Only a limited number of states have expanded privatization beyond allowing registered service companies to repair and return devices to commercial service without prior inspection by state inspectors. The inherent *conflict of interest* is often cited as the principal reason.

The NCWM Task Force on Planning for the 21st Century

The National Conference on Weights and Measures established a *Task Force on Planning for the Twenty-First Century* in 1990.[42]

> CHARGE: The Task Force on Planning for the 21st Century was appointed by Chairman N. David Smith in 1990 to assess the changes and impacts on weights and measures in the 21st Century. The Task Force was charged with:
>
> - *identifying issues which would change the nature of weights and measures;*
> - *reviewing possible strategies for addressing these issues; and*
> - *presenting recommendations to the NCWM Executive Committee for review and action.* [43]

After four meetings, the Task Force issued a Final Report in 1992. There were numerous recommendations but *privatization* of weights and measures regulation was identified as the highest priority. It was recommended that a comprehensive study be undertaken immediately.

> **_Conclusions_**
>
> The most pressing need at this time is to determine the limits of privatizing weights and measures service and regulation. There

is a strong perception that weights and measures is only a measurement service, that of device testing in order to determine if the device needs maintenance or repair. Our public leaders and businesses often do not grasp the regulatory aspects of weights and measures, that of providing a level playing field, a fair marketplace in which honest businesses can make a profit honestly, and in which consumers can buy a product or service by the amount that they are led to believe they are buying. In the sense that this government regulatory function is a service to the public, weights and measures is still a 'service.' The testing and repair of 100% of the population of devices in any jurisdiction can indeed be turned over to the private sector, but government must still monitor the honesty of businesses trading with the public and other businesses by weights or measures, and must then add monitoring of the private service agencies that do the majority of testing devices. The Task Force suspects that this will require more resources rather than less, with better trained government officials, with more sophisticated equipment to properly control the marketplace and keep it a fair and equitable environment for trade.

Recommendation: The Task Force believes that it should shift its focus from planning to an in-depth exploration of the issue of privatization. It the Conference does not immediately begin to define the limits of privatization, there will be fewer weights and measures regulatory programs in the future for which to plan![44]

The Final Report of the Privatization Work Group

A *Privatization Work Group* was established by the National Conference on Weights and Measures based on the recommendation of the Task Force on Planning for the 21st Century. The Privatization Work Group issued a Final Report in 1994.[45]

According to the Work Group, most states were already partially *privatized* to the extent that they permitted licensed private service agencies to place weighing and measuring devices back into service after repair or recalibration. However, there is much more to weights and measures regulation than simply devices. The Work Group pointed this out.

The Work Group found that those who had claimed to have privatized did not know the full extent of the weights and measures regulatory functions to maintain a fair marketplace. In general,

> governments that had 'privatized' had shifted device testing activities to private service and repair agencies, but had not found ways to privatize testing packaged commodities, transaction [pricing] verification, investigation of complaints, or the enforcement of weights and measures laws and regulations.[46]

The experiences of Kansas and Washington with privatization were examined by the Work Group:

Privatization of Weights and Measures in Kansas

Kansas has the most experience with privatization. Private service companies have been permitted to conduct annual inspections of commercial scales for three decades.

> For almost 30 years, the [D]epartment [of Agriculture] has used a unique, semi-privatized system in which it licenses technicians from private scale service companies to inspect commercial scales while a few state inspectors spot-check a fraction of the private work.[47]

State weights and measures inspectors conduct oversight of annual inspections by private service companies and assures that all commercial scales are tested by service companies.

> Kansas requires every commercial weighing or measuring device to be tested by a licensed service company of an authorized representative of a city or county with an established department of public inspection of weights and measures each year (excluding gas pumps which are regulated specifically by Weights and Measures personnel). Service companies and technicians must be licensed by the Weights and Measures program. Licensed companies and technicians are authorized to repair, install, and certify commercial weighing and measuring devices. Kansas is believed to be the only state that allows service technicians to certify commercial weighing and measuring devices.[48]

Kansas turned to privatization of commercial scale inspections because it did not have an adequate budget to hire enough inspectors to conduct annual inspections.

> Since Kansas had not the resources to inspect annually, they mandated that businesses buy annual inspections from the private

sector. However, they do not empower private companies to reject or condemn devices, and the State oversees the work of the private companies.[49]

Privatization of Weights and Measures in Washington

In the early 1990s, the State of Washington had a hybrid state-municipality weights and measures program. Over time, several cities had discontinued their weights and measures programs due to budget constraints (e.g., Tacoma). Only a few city programs remained: Seattle, Everett and Spokane (only Seattle remains today). State inspectors were scattered around the state and worked from home. Long travel times to inspections and re-inspections plus the difficulty in swapping equipment (e.g., LPG and loading rack meter portable test measures, heavy duty truck with vehicle scale test weights) among scattered inspectors made inspection efficiency problematic. State inspectors were burdened with even heavier inspection workloads as city programs shut down.

> [Washington]State's weights and measures resources has been allowed to erode so badly that there are some devices in rural areas that have not been tested in several years. The major population centers around Seattle and Spokane have city weights and measures agencies that cover their relatively small geographic areas. These agencies feel threatened by the State Legislature's linking a new fee for service program to what fees the cities can charge (most inadequate in urban areas) and when they can charge (only when a device passes inspection). The Washington Legislature also established a Task Force to study the current weights and measures situation and advise on implementing an appropriate system.[50]

Lobbyists for the regulated industry, especially gas station and food store trade associations, opposed any increase in funding for weights and measures enforcement. Instead, they sought privatization of weights and measures.

> The Washington State legislature was lobbied by trade groups that annual weights and measures government inspections duplicated private company testing: the legislature originally planned to give device owners the choice of using private companies or State agency. The program adopted by the State legislature provides for fees, but only when devices are found [not] in compliance.[51]

The State weights and measures code, Revised Code of Washington (RCW) Chapter 19.94, was the target of several proposed revisions. Some would have provided that all weights and measures programs in the State be supported by inspection fees and that only devices that failed inspection would be charged these fees. Fees would be set by an industry task force appointed by the Director of the Department of Agriculture. These bills did not pass because they were untenable. The fees were set very low so weights and measures inspectors would only be able to inspect devices located along the I-5 interstate highway corridor in order to limit travel time between inspections. The inspection fees would only support device inspections on a two-year frequency and there would be no funding for any package inspection activities or investigating consumer complaints.[52]

The Washington Office of Financial Management conducted a study on the cost-benefit of weights and measures inspections.

> The greatest risk weights and measures agencies face today is that governments at every level must reduce costs and when considering eliminating weights and measures agencies do not know of: (1) the regulatory responsibilities of weights and measures; (2) the cost/benefit of weights and measures programs; nor (3) any improvements that can reduce costs in regulatory efforts and increase benefits. The State of Washington's legislature had planned to eliminate weights and measures enforcement in that State. Instead, they agreed to fund a one-year study to determine whether weights and measures had substantial benefits as compared with the costs. ...Although one might question some aspects of the study, the benefits as measured were many times the costs. The question remaining from the study was not whether to retain the program, but how to pay for it.[53]

The Washington State Task Force on Weights and Measures published its conclusions. One conclusion was dismissive of the need for frequent inspections.

> The WA Task Force study concluded that there was no correlation between frequency of inspection and compliance.[54]

The NCWM Privatization Work Group explained the erroneous conclusion by the Task Force as being due to comparison of inspection data among states that used different standards. In other words, they were unwittingly comparing *apples and oranges*.

> Neither the effectiveness of a program itself nor the effectiveness of changes to a program can be tracked, estimated, or measured if the same standards are not used to evaluate the effects over time or in different locations. This was part of the problem encountered by the WA Task Force when they found widely varying compliance rates in different jurisdictions irrespective of the frequency of inspection.[55]

Economic Loss from Short Weight Packages

The NCWM Privatization Work Group stressed that the recording of actual errors 'as found' when conducting device inspections was critical in order to estimate the economic impact on consumers and producers.

> Work Group members who direct weights and measures regulatory programs submitted copies of their reporting forms and quickly came to the conclusion that standardized report forms are need in order to compare compliance data and error rates. The Work Group also noted that actual errors need to be recorded in order to make economic evaluations.[56]

The economic impact of short weight packages – perhaps due to not taking adequate tare or because a scale was out of calibration – was illustrated by this example.

> When weights and measures officials find short weight packages, many jurisdictions compute the economic loss in dollars using the shortage found, and apply that shortage to the amount of that product sold by a given retailer over some period of time. For example, if a shortage of 0.02 lb. is found for chicken breasts selling at $5.00 per pound, the shortage per package is $0.10. However, if the retailer sells 200 packages of chicken breasts per day, the economic shortage is $20 per day, $140 per week, $560 per month, and $6,720 per year for that one kind of package in that one store. This analysis is commonly done when determining whether to impose a fine, or take other legal action.[57]

Some Conclusions on Privatization

At present, the question of whether to privatize weights and measures is unresolved. Most states have taken the first step by allowing private service companies to place new and repaired devices in service without an inspection by state weights and measures inspectors. This was a practical step because state inspectors were increasingly stretched thin, with too many devices and too large a coverage area, and unable to schedule a timely visit. However, it is very clear that there is always a conflict of interest in allowing a for-profit company to test, repair and replace devices – they are not a disinterested third party like state and local weights and measures inspectors who are government officials and bound by strict ethics rules. States usually have audit programs for random checks of work by private service companies but this is often spotty due to a lack of inspector resources.

Few states have taken next steps toward increased privatization. Service agents cannot perform enforcement work for violations of weights and measures laws (e.g., enter premises, collect evidence, take statements, cite violations, testify in court proceedings) or investigate consumer complaints because they have no special police commission. The experience, training, knowledge, equipment, and references (e.g., NIST handbooks, weights and measures laws and regulations) of individual service agents are often unsatisfactory. Perhaps, even more important, service agents maintain devices and cannot help with packaged goods weights and measures work such as price scanning, checking net contents of packaged goods, unit pricing, method of sale and labeling.

The annual cost of weights and measures inspections is minimal so privatization saves little money for a business. For example, the annual registration fee for a retail-motor-fuel dispenser in Washington is $10. A typical 24-pump gas station would pay $240 annually or about $0.66 per day. Replacing weights and measures inspectors with service agents merely shifts this cost but does not remove it.

Privatization may save the government money – although many weights and measures programs are largely self-funded by registration and inspection fees – but the focus is solely on businesses. Importantly weights and measures regulation has two roles: (1) promoting a level playing field, or fair competition, among competing businesses and (2) consumer protection or making sure that consumers get what they pay for. State weights and measures programs are often more responsive to business lobbying because industry associations typically target lobbying at state legislatures. Unlike industry trade groups, there are few organizations that represent the consumer. Weights and measures regulatory programs are the principal advocate for consumers. State legislatures are more likely to be split between the major political parties. One party, more dominant in small cities and rural areas, consistently opposes any fee increases believing that, *a fee is a tax by another name*.

On the contrary, councilmembers for large cities and populous counties tend to emphasize consumer protection in recognition of the fact that consumers are also voters.

Notes

1 Federal Trade Commission. "Milk: Does It Measure Up?: A Report on the Accuracy of Net Contents Labeling of Milk and Other Products" (July 17, 1997), p. 1.
2 Ibid., Appendix A, Table 1, p. 7.
3 Ibid., Tables 2–3, pp. 7–8.
4 National Institute of Standards and Technology. Special Publication 1053. *Report of the 91st National Conference on Weights and Measures*. (2006), p. GS-2. "President's Address to the National Conference on Weights and Measures at Chicago, Illinois on July 11, 2006 by Dr. William A. Jeffrey, NIST Director."
5 U.S. Department of Agriculture. "Milk Production" (February 20, 2020), Table "Milk Cows and Production – United States: 2010–2019, p. 7.
6 U.S. Department of Agriculture. "Retail Milk Prices Report" (December 23, 2019), p. 1.
7 "Milk: Does It Measure Up?", op. cit., p. 6.
8 Ibid., p. 1.
9 Ibid., Note 8, p. 17. As the manager of the Consumer Affairs Unit in Seattle, was responsible for five regulatory programs: weights and measures, taxicabs and for-hire vehicles, limousines, transportation network companies (e.g., Uber, Lyft) and towing. I was employed in this role for more than 20 years (1996–2017).
10 National Institute of Standards and Technology. Handbook 133 *Checking the Net Contents of Packaged Goods* (2020), Appendix C, "Standard Package Report," pp. 181–82.
11 Ibid., Note 12, p. 17
12 National Institute of Standards and Technology. Special Publication 942, *Report of the 84th National Conference on Weights and Measures* (1999), Board of Directors, Appendix F, pp. BOD-55 – BOD-57.
13 National Institute of Standards and Technology. Special Publication 1053, Report of the 91st National Conference on Weights and Measures (2006), Board of Directors, pp. BOD-2 – BOD-3.
14 Judy Cardin, Chief of Wisconsin Weights and Measures. "Conducting Effective Marketplace Surveys and National Investigations: 2010 Multi-State Seafood Investigation." Presentation at the National Conference on Weights and Measures annual meeting in Portland, Maine (July 2012). The states participating were: Alaska, California, Colorado, Connecticut, Florida, Illinois, Iowa, Kansas, Maine, Michigan, Minnesota, Missouri, North Carolina, New York, Ohio, Washington and Wisconsin.
15 Ibid.

16 Ibid.
17 NBC *Dateline*, "Highway Robbery", January 20, 1999.
18 Interview with Kurt Floren, op. cit.
19 Juana Williams. "Fuel Tank Capacity and Gas Pump Accuracy" *Weights and Measures Quarterly*. Vol. 8, No. 3 (August 2005), pp. 8–9.
20 Kurt Floren, Agricultural Commissioner/Director of Weights and Measures for the County of Los Angeles. "Retail Petroleum Fraud Investigation: Brief Summary" (1999), p.3 of 4.
21 Interview with Kurt Floren, Agricultural Commissioner/Director of Weights and Measures for the County of Los Angeles, conducted by video-conference on April 6, 2021.
22 Floren. "Retail Petroleum Fraud Investigation: Brief Summary", op. cit., p. 1 of 4.
23 Ibid., p. 3 of 4.
24 Floren. "Retail Petroleum Fraud Investigation: Brief Summary", op. cit., p. 1 of 4.
25 Ibid.
26 Interview with Kurt Floren, op. cit.
27 Ibid., pp. 2–3 of 4.
28 Floren. "Retail Petroleum Fraud Investigation: Brief Summary", op. cit., p. 4 of 4.
29 Claire Vitucci. "Prosecutors: High-tech gas pump fraud bilked consumers out of $1 million" *The Daily Transcript. (San Diego's Business Daily)*, October 9, 1998
30 NBC *Dateline*, "Highway Robbery", January 20, 1999.
31 *Los Angeles Almanac*. "Gasoline Prices Monthly & Annual Averages: 1978 through 2021." http://www.laalmanac.com/energy/en12.php Accessed April 7, 2021.
32 National Association of Convenience Store (NACS). "Key Facts About Fueling" www.convenience.org Accessed April 7, 2021.
33 Interview with Kurt Floren, op. cit.
34 See Chapter 4, Tables 4.2 and 4.3.
35 Randy Awdish. "The Following Will Surprise or Even Shock You: A Look into TV Newsmagazines and Their Effect on Consumers" *Loyola Consumer Law Review*, Volume 13, Number 4 (2001), p. 363.
36 Ibid.
37 Ibid., p. 368
38 NBC *Dateline*, "Highway Robbery", January 20, 1999.
39 National Institute of Standards and Technology. Handbook 155. *Weights and Measures Program Requirements: A Handbook for the Weights and Measures Administrator*. (2011), p.64.
40 National Institute of Standards and Technology. Special Publication 854. Report of the 78th *National Conference on Weights and Measures. Kansas City, Missouri* (1993) Report of the Executive Committee, Appendix G "Privatization Work Group: October 9–10, Richmond, VA" p. 126.
41 Ibid., p. 69.

42 National Institute of Standards and Technology. Special Publication 793. *Report of the 75th National Conference on Weights and Measures*. Washington, D.C. (1990). Report of the Executive Committee, p. 38.
43 National Institute of Standard and Technology. Special Publication 845. *Report of the 77th National Conference on Weights and Measures*. Nashville, Tennessee (1992) Report of the Executive Committee, Appendix E, p. 116.
44 Ibid., pp. 124–125.
45 National Institute of Standards and Technology. Special Publication 870. *Report of the 79th National Conference on Weights and Measures.* San Diego, California (1994). Report of the Executive Committee, Appendix A, pp. 69–113.
46 Ibid., p. 69.
47 Andy Marso. "State's scales tipping more toward accuracy" *The Topeka Capital-Journal* (cjonline.com). August 18, 2014.
48 Kansas legislative Research Department. "Weights and Measures Program" (September 20, 2018), p.1/5.
49 NIST Special Publication 870, op. cit., Appendix A, p. 71.
50 NIST Special Publication 854, op. cit., Report of the Executive Committee, Appendix G "Privatization Work Group." p. 138
51 NIST Special Publication 854. op. cit., Report of the Executive Committee, Appendix G "Privatization Work Group." p. 127.
52 See, for example, State of Washington HB 2522 and companion bill SB 6099 (1994).
53 NIST Special Publication 854. op. cit., Report of the Executive Committee, Item 101–9 "Organization: Privatization Work Group" p. 60.
54 Ibid., Allan M. Nelson. "Report of the Privatization Work Group meeting, November 4–6, 1993, Seattle, WA." p. 104.
55 Ibid., p. 109.
56 Ibid., p. 71.
57 Ibid., p. 110.

Bibliography

Awdish, Randy. "The Following Will Surprise or Even Shock You: A Look into TV Newsmagazines and Their Effect on Consumers" *Loyola Consumer Law Review*, Volume 13, Number 4 (2001), pp. 353–379.

Federal Trade Commission. "Milk: Does It Measure Up?: A Report on the Accuracy of Net Contents Labeling of Milk and Other Products" (July 17, 1997).

National Institute of Standards and Technology. Special Publication 793. *Report of the 75th National Conference on Weights and Measures*. Washington, D.C.: National Institute of Standards and Technology (1990).

National Institute of Standard and Technology. Special Publication 845. *Report of the 77th National Conference on Weights and Measures*. Nashville, Tennessee: National Institute of Standards and Technology (1992).

National Institute of Standards and Technology. Special Publication 854. *Report of the 78th National Conference on Weights and Measures.* Kansas City, Missouri: National Institute of Standards and Technology (1993).

National Institute of Standards and Technology. Special Publication 870. *Report of the 79th National Conference on Weights and Measures.* San Diego, California: National Institute of Standards and Technology (1994).

National Institute of Standards and Technology. Special Publication 942, *Report of the 84th National Conference on Weights and Measures.* National Institute of Standards and Technology (1999).

National Institute of Standards and Technology. Special Publication 1053. *Report of the 91st National Conference on Weights and Measures.* National Institute of Standards and Technology (2006).

National Institute of Standards and Technology. Handbook 155. *Weights and Measures Program Requirements: A Handbook for the Weights and Measures Administrator.* National Institute of Standards and Technology (2011).

National Institute of Standards and Technology. Handbook 133. *Checking the Net Contents of Packaged Goods.* National Institute of Standards and Technology (2020).

NBC Dateline. “Highway Robbery” January 20, 1999.

6 Landmark Legal Cases

Several legal decisions have had a major impact on how weights and measures regulation is conducted in the United States. Among those cases, the issue of '*moisture loss*' (formerly, '*shrinkage*' *or* '*evaporation*') was, and still remains, the most controversial. This chapter examines the legal cases pertaining to *net contents* of packaged goods.

The concept of '*shrinkage*' is, at once, both simple and complex. It refers to the fact that some packaged goods meet the declaration of 'net contents' on the package label when the packages are filled at the point-of-pack (production), but not later, when the package is offered to the consumer at the point-of-sale (retail). The loss of net contents is normally blamed, by the packager, on factors not under its direct control such as transportation, handling and storage methods. Much of the early debate on 'shrinkage' centered on whether there was actually any product loss or whether packages were short measure to begin with when they left the packager. Subsequently, the debate focused on how much 'shrinkage' there was for a specific commodity, if any, and whether the packaging was unsatisfactory. However, most weights and measures officials rejected the 'shrinkage' argument outright, stating that it missed the point entirely. These weights and measures officials strongly argued that it was the sole responsibility of the packager to select and fill the packages in such a manner that the consumer received the full 'net contents' specified on the package label at the time the package was purchased. They reasoned, 'what good is the declaration of net contents if it wasn't true?'

Washington State Supreme Court 'Shrinkage' Case

The Washington State Supreme Court decision in the *City of Seattle vs. Goldsmith* (1913) found in favor of the consumer, confirming the principle that 'net contents' meant '*net contents at time of sale.*' This

DOI: 10.4324/9781003263661-7

appeared to have settled the issue, at least in Washington, but that was not to be. Almost immediately, the food industry lobbied federal agencies with jurisdiction over packaging, under the Pure Food and Drug Act (1906). The Gould Amendment (1913) allowed for '*reasonable variations*' from labeled net contents on packaged goods:

> Be it enacted by the Senate and House of Representatives of the United States of America in Congress assembled, That section eight of an Act entitled "An Act for preventing the manufacture, sale, or transportation of adulterated or misbranded or poisonous or deleterious foods, drugs, medicines, and liquors, and for regulating traffic therein, and for other purposes," approved June thirtieth, nineteen hundred and six be, and the same is hereby amended by striking out the word, "Third. If in package form, and the contents are stated in terms of weight or measure, they are not plainly and correctly stated on the outside of the package," and inserting in lieu thereof the following:
>
> "Third. If in package form, the quantity of the contents be not plainly and conspicuously marked on the outside of the package in terms of weight, measure, or numerical count: Provided, however, **That reasonable variations shall be permitted,** and tolerances and also exemptions as to small packages shall be established by rules and regulations made in accordance with the provisions of Section three of this Act."[1] [**bold** print added for emphasis]

Nearly 64 years later, the matter would finally be settled by the U.S. Supreme Court decision in *Jones v. Rath Packing Co.* (1977).

Washington State Supreme Court 'Shrinkage' Case (1913)

Seattle v. Goldsmith
73 Wash. 54 (April 15, 1913)
Appellant: The City of Seattle
Respondent: J. S. Goldsmith

The Washington State Supreme Court reversed a judgment by the Superior Court for King County dated February 18, 1912 in a landmark legal decision in favor of the City of Seattle in the case of *City of Seattle vs Goldsmith* on April 15, 1913. The case concerned the Seattle municipal ordinance requiring that packages must meet the labeled net contents (weight of product only, excluding tare). This decision established a legal precedent in Washington that the packer was responsible to ensure that the net contents be accurate at the point-of-sale and not only at the point-of-pack. Later, this principle became known as '*net contents at time of sale.*'

> The defendant contended that [packages of raisins, salt and pails of lard] were subject to *shrinkage* and therefore it would be unreasonable and unjust to require the contents to be stated since the markings might be incorrect at the time the package reached the consumer although correct at time packed. The Court held that if a commodity is subject to shrinkage no one is in a better position to know that fact than the packer and it is his duty to make allowance for same so that **when the package reaches the consumer it shall be full weight**. The item of lard was involved in two other cases brought into court at the same time as the Goldsmith case but by agreement it was decided that the one decision should govern the three cases.[2]

The main argument of the respondent, Goldsmith, was that the Seattle ordinance regarding 'net weight' was unreasonable in that it did not make any allowance for loss of product weight by '*evaporation*.' The Court disagreed.

> The next contention of respondent goes to the reasonableness of the ordinance in failing to make allowances for the **loss of weight by evaporation**, stating in support of his theory that California salt packed in sacks, and raisins packed in cartons, the two commodities embraced in the complaint, will lose weight by evaporation., and that the true weight if stated on the container at the time of packing would not be the true weight at the time of delivery to the consumer, and hence the dealer would be liable for a violation of the Federal act requiring that, if the weight be stamped on the package, it must be the true weight. This would mean, assuming respondent's contention as to loss of weight by evaporation to be true, that the loss must fall on the consumer. It does not appear to us that a law is unreasonable because compliance with its requirements shifts this loss to the original packer or manufacturer. It is not unreasonable to require that the packer or manufacturer shall ascertain this loss by evaporation as he is best in position to do, and overcome the loss by increasing the size of the package or the weights of the commodity packed therein, or withhold his goods from the market until it is possible to ascertain the true net weight. Whatever may be the necessary course to adopt to enable the container to correctly indicate the weight of the commodity it contains, it is not unreasonable

> to place that burden upon the one who puts the article before the public as a sale commodity, and compel him, if he wishes to retain his trade, to so pack his commodities that the consumer may know the true quantity of the thing he buys, and thus protect himself in paying the value of the thing he buys. At all events we apprehend that there will be little likelihood of the honest merchant subjecting himself to a penalty under the ordinance if he is able to show that, in an honest endeavor to comply therewith, the nature of the article is such that an absolute compliance with its terms is impossible. The power of the city to pass the ordinance being sustained, it will be a simple matter to so amend it, if found to be necessary, as to conform to all natural and uncontrollable conditions.[3]

Over the years, the term "*reasonable variations*," as used in the Gould Amendment, was not defined by federal agencies responsible for implementing the Pure Food and Drug Act or any related laws addressing package labeling. This rendered enforcement, by state and local weights and measures regulatory programs, problematic.

> ...for many years, at both the FDA, the United States Department of Agriculture (USDA) and the state level, the principle of loss of weight due to **loss of moisture** has been recognized. It also should be noted that, though the[4]principle was recognized, no attempt has been made to spell out the allowable variation on a product-by-product or package-by-package basis. The variation has been informally allowed on a 1 or 2 percent basis.[5]

California 'Moisture Loss' Cases

California state weights and measures jurisdictions, employing state package net contents laws and rules, came into conflict with federal packaging laws and rules in the early 1970s regarding prepackaged meat. Specifically, weights and measures inspectors in the counties of Los Angeles (Becker) and Riverside (Jones) found packages of Rath bacon short weight and ordered them off sale.

> During the period September 1971 to March 1972 inspectors under the supervision of Becker and Jones visited supermarkets in Los Angeles and Riverside Counties and weighed packages of Rath bacon to determine compliance with the State statute and regulations concerning net weight labeling. Becker's representatives ordered approximately 84 lots of bacon off sale for short weight;

> Jones ordered nearly 400 packages of Rath bacon off sale in the period September 29 to December 30, 1971, for the same reason.[6]

In addition to ordering underweight product off sale, Los Angeles and Riverside counties sought civil penalties from Rath Packing Company for false advertising.

> Damages of $2,500 per time each short-weight package of bacon were asked. This amounted to $2,900,000 in Los Angeles and $800,000 in Riverside.[7]

Subsequently, Rath Packing Company sought an injunction to prevent these enforcement actions, claiming that federal law preempted state law with regard to labeling net contents of packaged meats subject to inspection by the U.S. Department of Agriculture pursuant to the federal Wholesome Meat Act of 1967.

U.S. District Court, Central District of California. (1973)

Rath Packing Co. v. Becker
357 F. Supp. 529 (C.D. Cal., 1973)
Plaintiff: Rath Packing Company

Defendants:
M.H. Becker, Director of Los Angeles County, Department of Weights and Measures

Joseph W. Jones, Director of Riverside Country, Department of Weights and Measures

C. B. Christensen, California Director of Agriculture *(Intervenor)*

The complaints filed by Los Angeles County (Becker) and Riverside County (Jones) were consolidated. Each had found Rath packaged meat products to be short-weight and ordered them off-sale pursuant to the California Business and Professions Code, Sec. 12211 (statute) and Title 4, California Administrative Code, Ch. 8, subch. 2, Art. 5. (regulation). In dispute were the relevant provisions of the federal Wholesome Meat Act of 1967 at 21 U. S. Code Sec. 601(n)(5) which included an exception in the definition of "misbranding" (mislabeling) to allow "reasonable variations." 21 U.S. Code Sec. 679 preempts states from adopting requirements "in addition to or different than" the federal requirement.[8]

The rule (regulation) issued by the Secretary of Agriculture pursuant to the Wholesome Meat Act of 1967 elaborated on "reasonable variations" at Title 9 Code of Federal Regulations (CFR) Sec. 317.2(h)(2) as follows:

> (2) The statement as it is shown on a label shall not be false or misleading and shall express an accurate statement of the quantity of contents of the container exclusive of wrappers and packing substances. Reasonable variations caused by loss or gain of moisture during the course of good distribution practices or by unavoidable deviations in good manufacturing practice will be recognized. Variations from stated quantity of contents shall not be unreasonably large.

The defendants argued that this rule was void due to vagueness but the court ruled that California could not substitute its own standard even though the U. S. Department of Agriculture had not specified the amount of moisture loss that would not be unreasonable.

Since the California rule for determining 'net contents' did not allow for moisture loss it was found to be preempted by federal law and could not be enforced against Rath meat products subject to inspection under the federal Wholesome Meat Act of 1967.

U.S. Court of Appeals, Ninth Circuit (1976)

Rath Packing Co. v. Becker
530 F. 2d 1295 (9th Cir. 1976)
Plaintiff: The Rath Packing Company

Defendants:
M.H. Becker, Director of Los Angeles County, Department of Weights and Measures

Joseph W. Jones, Director of Riverside Country, Department of Weights and Measures

C. B. Christensen, California Director of Agriculture (Intervenor)

The U. S. Court of Appeals reviewed the district court holdings in Rath Packing Co. v. Becker and found as follows:

(1) that the district court had jurisdiction over the subject matter of this case, personal jurisdiction being conceded;
(2) that the district court erred in invalidating 9 CFR 317.2(h)(2);
(3) that the Wholesome Meat Act of 1967, 21 U.S.C. § 601 et seq., and 9 CFR 317.2(h)(2) preempt Cal.Bus. and Prof.Code § 12211

and 4 Cal.Admin. Code ch. 8, subch. 2, Art. 5, and that Becker, Jones, and Christensen were properly enjoined from enforcing those sections;

(4) that the district court correctly held that state standards not in addition to or different than the federal net weight labeling standard may be enforced by appropriate State procedures at the retail level; and

(5) that 4 Cal.Admin.Code ch. 8, subch. 2, Art. 5.1, is preempted by federal law, that Cal.Bus. and Prof.Code § 12607 is preempted by federal law to the extent indicated in part V, supra, and that their enforcement should be enjoined.

Accordingly, the judgment of the district court is affirmed in part, reversed in part, and the case is remanded for entry of an amended order in conformance with this opinion.[9]

U. S. Supreme Court 'Moisture Loss' Case (1977)

Jones v. Rath Packing
430 U. S. 519 (1977)

Petitioner:
Joseph W. JONES, Director of the County of Riverside, California, Department of Weights and Measures

Parties: The Rath Packing Company et al.

The Supreme Court reviewed the holdings by the District Court and the Court of Appeals and held:

> With respect to respondent packing company's packaged bacon, Sec. 12211 and Art. 5 are pre-empted by the FMIA [Federal Meat Inspection Act as amended by the Wholesome Meat Act]. Since California makes no allowance for loss of weights resulting from moisture loss during the course of good distribution practice the state laws' requirement that the label accurately state the net weight, with implicit allowance only for reasonable manufacturing variations is 'different than' the federal requirement which permits manufacturing deviations and variations caused by moisture loss during good distribution practice.[10]

Rath produced packages of bacon at plants inspected by the U.S. Department of Agriculture pursuant to the Federal Meat Inspection Act (FMIA). These inspections included verifying labeled net weight.

States are granted concurrent jurisdiction to verify net contents after the packages leave the packaging plants for distribution to retail stores. However, the states are pre-empted from applying different standards of net contents that don't include an allowance for reasonable variations due to moisture loss. Even though the Secretary of the Department of Agriculture has not specified moisture loss amounts that would be 'reasonable,' the state of California is not permitted to substitute its own definition because it has no allowance for moisture loss.

During the previous cases, Rath Packing Company provided some specific information about moisture loss from the packages of bacon in question.

> n. 10. After it is packed, bacon loses moisture. Some of that moisture is absorbed by the insert on which the bacon is placed. A wax board insert will absorb approximately 5/16 of an ounce from the product, whereas a polyethylene insert will absorb approximately 1/16 of an ounce. App. 88–90, 94; 530 F.2d, at 1299 n. 2. In addition, moisture is lost to the atmosphere or, in a hermetically sealed package, by condensation onto the packing material. App. 61. California's inspectors include in the weight of the material any moisture or grease which the bacon has lost to it. Federal inspectors at the packing plant, by contrast, determine the tare by weighing the packing material dry. 530 F.2d at 1299. It is not feasible for field inspectors to use a dry tare method. C. Brickenhamp, S. Hasko, & M. Natrella, Checking Prepackaged Commodities Revision of National Bureau of Standards Handbook 67, p. 33 (July 1975 Draft).[11]
>
> n. 16. Both sources of variation from stated weight are relevant to bacon. Bacon loses moisture to its wrapping materials and to the atmosphere. See n. 10, supra. The rate of loss to the atmosphere in a typical retail showcase is 0.3/16 to 0.4/16 of an ounce per day. App. 95. In addition, since bacon is cut in discrete slices, it is impossible to guarantee that each package will contain exactly the stated weight when packed. Instead of seeking exactitude, Rath approved packages if they were within 5/16 of an ounce of a target weight. Prior to petitioner's enforcement activities, and the similar activities of Becker, see n. 4, supra, Rath's target weight was 3/16 of an ounce over the stated weight, or 1 lb. 3/16 oz. for a one-pound package. Thus, a package would be passed if it weighed between 15 14/16 oz. and 1 lb. 8/16 oz. In response to the California enforcement measures, Rath raised its target weight to 8/16 oz. over stated net weight for bacon packed on a polyethylene insert, and 12/16 oz. over stated weight of bacon packed on wax boards. App. 86–89.[12]

Wet Tare v. Dry Tare

The preceding discussions in the California 'moisture loss' cases introduced the terms 'wet tare' and 'dry tare.' Previous to the Supreme Court decision, '*wet tare*' was the method used by weights and measures inspectors to verify 'net contents' of packaged goods. After the decision, 'wet tare' is no longer used for meat products subject to inspection by the U. S. Department of Agriculture. Instead, '*unused dry tare*' is used to check net contents at point-of-pack and '*used dry tare*' is used to check net contents at wholesale and retail.

'Wet tare' refers to a 'net weight' inspection procedure where the meat is removed from the packaging and weighed. Moisture that evaporated and condensed on the inside of the packaging materials and liquid (e.g., purge) that leaked from the meat into the soaker pads and tray are considered 'tare' and are excluded from 'net weight.'

When meat products are packaged, tare materials (wrappings, tray, soaker pads) are 'unused.' Net weight verification determinations exclude 'unused dry tare.' Once a packaged meat product is placed in the refrigerated display cases and offered for purchase by the consumer, the tare is in 'used' condition. Net weight verification methods used at retail require an inspector to remove liquid from tare materials by drying before subtracting 'used dry tare' from 'gross weight' to determine 'net weight.'

Moisture Loss Determinations

Moisture loss allowances have been determined for some product categories and published in NIST Handbook 133 *Checking the Net Contents of Packaged Goods*. For example, Table 2-3 "Moisture Allowances" specifies that the moisture allowance or so-called '*gray area*' for flour is 3 percent and for dry pet food is 3 percent.[13] The step-by-step method of determining the magnitude of permitted variations for packaged products is described in NIST Handbook 133 also.

Closely related to net contents determinations is the issue of ice glaze tare on seafood and poultry. The ice glaze can constitute a significant portion of gross weight for those products and especially seafood. NIST Handbook 133 provides the accepted procedure for removing ice glaze at Section 2.6.2.2 "Net Weights of Ice Glazed Seafood, Meat, Poultry or Similar Products." For example, sellers or inspectors are instructed to remove glaze using cold water to gently melt the ice while the product is in a sieve inclined at 17–20 degrees (for uniform draining) and supported by a tilt block for a period of two minutes. The product is to remain rigid, not thawed.[14] Many weights and measures jurisdictions have adopted the most recent edition of NIST Handbook 133 into law by reference.

Notes

1 *The Pure Food and Drug Act* (1906) as revised by Public Law 672–419 (1913), known as the Gould Amendment. The definition of "misbranded" packages was changed to provide for "reasonable variations" from net weight.
2 City of Seattle. *Annual Report of the Department of Public Utilities, 1913*, p. 96. Also, see *Proceedings of the Annual Convention of the League of Washington Municipalities* at Spokane, Washington November 19–22, 1913. Sixth Session: "Department of Weights and Measures" by A. W. Rinehart, November 21, 1913, pp. 95–97. **My emphasis**.
3 *Cases Determined in the Supreme Court of Washington*, Vol. 73, pp. 61–62. "Seattle v. Goldsmith" (April 1913). **My emphasis**.
4 See 9 CFR 317.2(h)(2).
5 Harvey L. Hensel. "Look What Consumerism Has Done Now" *Food, Drug, Cosmetic Law Journal* Vol. 29, No. 5 (May 1974), p. 221.
6 Rath Packing Company v. Becker 530 F. 2d 1295:: CaseText, "Procedural History."
7 Hensel, op. cit., p. 222.
8 Rath Packing Company v. Becker 357 F. Supp. 529 (C.D. Cal., 1973):: Justia.
9 Rath Packing Company v. Becker 530 F. 2d 1295:: CaseText, "Conclusion."
10 430 U.S. 519 (1977)
11 Jones v. Rath Packing, 430 U.S. 519 (1977)::law.resources.org, Note 10.
12 Ibid., Note 16.
13 National Institute of Standards and Technology. Handbook 133 *Checking the Net Contents of Packaged Goods* (2020), p. 27.
14 Ibid., pp. 36–37.

Bibliography

City of Seattle. *Annual Report of the Department of Public Utilities, 1913.*

National Institute of Standards and Technology. Handbook 133. *Checking the Net Contents of Packaged Goods*. National Institute of Standards and Technology (2020).

League of Washington Municipalities of Spokane. *Proceedings of the Annual Convention of the League of Washington Municipalities* at Spokane, Washington, November 19–22, 1913.

7 Conclusions and Forecasts

The foregoing chapters have examined the origins, history and current state of weights and measures regulation. This last chapter will attempt to forecast the future direction that weights and measures regulation may take. But first, let's begin with a summary of the present state of weights and measures regulation in the United States.

The Current State of Weights and Measures Regulation

State and local government (county, municipal) weights and measures regulatory jurisdictions are tasked with, among other things, verifying that weighing and measuring devices are accurate and that the labeled net contents of packages are correct. The focus is on assuring accurate *quantity* in marketplace transactions. This really amounts to verifying that prices are accurate for the quantity of a product delivered. In other words, short-measure transactions are actually an example of overcharging the consumer.

How well state and local weights and measures jurisdictions are regulating the marketplace to achieve the twin goals of consumer protection (assuring that buyers get what they pay for) and equity in the marketplace (providing a level playing field for sellers) is *not* a settled question. Why is it so difficult to measure the effectiveness of weights and measures regulation? Let's summarize some of the issues raised in the preceding chapters that bear on the effectiveness of weights and measures regulation.

Effectiveness of Weights and Measures Regulation

Measures of Effectiveness. Unfortunately, it is very difficult, and perhaps impossible, for many state weights and measures officials to

DOI: 10.4324/9781003263661-8

properly evaluate the effectiveness of their regulatory programs. This state of affairs has several and diverse causes: (1) One major obstacle to computing *measures of effectiveness* (MOE) is that there is often a lack of *reliable* weights and measures device inspection data and package inspection data. Without standardized device categories as a common denominator, analysts are attempting to add 'apples and oranges.' This was the main difficulty in conducting the 2019 survey of state weights and measures programs – just as it was with the 2002 survey. (2) Another significant impediment to developing measures of effectiveness was the absence of complete raw data to examine. For instance, some states don't even maintain device inventories so they don't know how many weighing and measuring devices they are supposed to inspect. This is most common in states that don't assess annual device registration fees. (3) In states that have embraced partial privatization, poor recordkeeping by private service companies often means that 'as found' device calibration errors are unknown. Without 'as found' data, it is virtually impossible to estimate overcharges that have been prevented. This, in turn, frustrates any meaningful cost: benefit analysis. (4) Too frequently, inspection databases are obsolete because budget reductions defer any upgrades. As a result, management cannot utilize *feedback loops* to assess what changes may be needed to improve weights and measures programs (e.g., staffing, training, planning, inspections, data collection, measures of effectiveness) and make them both more efficient and more effective.

Partial Privatization. No weights and measures program is funded to perform full oversight over the regulated industry. As a result, programs must prioritize inspections where they will yield the largest payoff, that is, where there is the most serious noncompliance (e.g., largest number of consumers impacted, highest failure rates, largest calibration errors). In order to free up inspector resources several steps have been taken: (1) Inspection procedures are frequently abbreviated to speed up inspections. For example, it is increasingly common for inspections to eliminate slow fill tests of retail motor-fuel dispensers. (2) Rather than inspect every fuel dispenser every year, some jurisdictions only have the resources to inspect a sample. (3) State and local weights and measures regulatory programs have had to reduce device and package inspection frequencies across the board due to the tension between growing device inventories and shrinking budgets. Because of this, most states have turned to partial privatization to help. However, experience has shown that privatization of inspection activities, by substitution of private service companies for weights and measures inspectors, necessarily involves an economic *conflict of interest* that is distinctly at odds with the goal of consumer protection.

General Fund Monies. Without General Fund monies in addition to regulatory fee revenue, such as registration/license fees or inspection fees, many weights and measures jurisdictions simply do not have adequate budgets to conduct package inspection activities. Also, device registration fees cannot be used to subsidize package inspections. The package inspection activities that are forgone include checking net contents of packaged goods, verifying pricing accuracy in electronic scanning systems, inspecting labels, checking method of sale, and verifying unit price code compliance. Even in weights and measures jurisdictions where some package net contents inspections are conducted, gross weight surveys are often substituted for the more time-consuming full Handbook 133 inspections. In the marketplace, goods in packaged form are quickly replacing goods purchased from bulk over a scale or through a meter, so this trend of ignoring package inspection activities is increasingly problematic. To illustrate, the number of scales was nearly unchanged (+0.7%) between 2002 (1,0109,155) and 2019 (1,017,217), even though the population increased 13.5 percent (from 288,600,000 to 327,533,795) during the same period, according to state surveys of weighing and measuring devices.[1] This means that the number of scales may have already peaked.

Optimum Device Inspection Frequency. Many state weights and measures programs do not prepare annual reports assessing the effectiveness of their device and package inspection programs in improving consumer or producer protection. However, there is normally some attempt to track consumer complaint trends and failed inspections as a percentage of all inspections. Complaints are investigated but device owners are usually not targeted based on their compliance histories even though, unquestionably, there is an inverse relationship between inspection frequency and inspection failure rate. This is demonstrated by Figure 1-1, a graph of retail motor-fuel dispenser inspection frequency v. inspection failure rates for 19 states that participated in the 2019 survey of state weights and measures programs. The longer the period between inspections, the higher the failure rate. However, no studies have been performed to compute an *optimum* inspection frequency balancing consumer protection with effective allocation of limited inspector resources. At some point, there are diminishing returns from increasing inspection frequency. Despite this, shrinking budgets, in the face of growing inspection workloads have reduced both the thoroughness and frequency of inspections.

Forecast: The Future of Weights and Measures Regulation

Regulatory Fees. The future of weights and measures regulation depends considerably on the growth of the general economy because that has a direct impact on government tax revenue and, therefore, operating budgets for government agencies. The transition away from the General Fund to regulatory fees may help stabilize funding for weights and measures regulatory programs. However, weights and measures regulatory programs always find it difficult to raise regulatory fees due to predictable opposition from industry lobbyists. The population has been growing rapidly and this generally means a proportionate increase in weighing and measuring devices. The count of retail motor-fuel dispensers in the United States increased by 16.9 percent between 2002 and 2019 while the population grew 13.5 percent during the same 17-year period.[2] But, the increase in the number of commercial scales was negligible. As a result, in the long run, the reliability of device registration fees as a predictable source of funding for weights and measures regulatory programs is questionable.

Internet Sales. During the COVID-19 pandemic (2020–2021), most people shopped online for many consumer goods including groceries. This magnified potential issues related to package inspections since many items were shipped directly to the consumer from other states or other countries that may not conduct routine package inspection activities. Investigation of consumer complaints is problematic since, as everyone who has shopped online knows, it is nearly impossible to contact sellers – or even determine which state they are located in. Several years ago, some weights and measures jurisdictions would schedule a periodic *surf day*, normally before holidays, to search the internet (*surf the net*) for package violations such as 'method of sale' (e.g., *large* frozen turkey instead of price per pound). Fewer than one-half of online sellers posted contact information a decade ago so it was not possible to alert the weights and measures regulatory program with legal jurisdiction to follow up. Many sellers were not even located in the United States. In the future, this situation will continue to worsen.

New Technology. Technological innovations change the nature of weights and measures, often profoundly and in unexpected ways. This greatly complicates inspection work. For example, the development of alternative fuels such as compressed natural gas (CNG), hydrogen fuel,[3] and electric charging stations[4] have all presented challenges for developing new inspection procedures, designing test equipment and training inspectors. As gasoline-powered vehicles are replaced by electric vehicles, fuel dispenser registration fees – the source of more than one-half of all regulatory fee revenue – will also disappear and create a huge funding crisis for the majority of state and local weights and measures regulatory programs. Retail motor-fuel dispensers and commercial scales represent about 90 percent of all weighing and measuring devices. Therefore, they are the principal source of fee revenue that support weights and measures regulation. Electric vehicles are already starting to create funding concerns for state highway maintenance and construction because those vehicles are not subject to state or federal excise taxes added to fuel prices. The reluctance of legislators to deal with this emerging problem, for many years now, is indicative of how difficult it will also be for weights and measures programs to replace the device registration fee revenue. Since several major automakers are committed to ceasing the manufacture of most models of gasoline-fuel vehicles as soon as the next 20 years, the fuel dispenser registration fee revenue issue may become a serious, almost existential, problem for weights and measures regulatory programs within a decade.

Sophisticated Electronic Fraud. Fraud prevention at fuel dispensers has added a new dimension to retail motor-fuel dispenser inspections. Here are some examples: (1) Gas stations with large-error short-measure complaints could be part of a fraud scheme that can only be detected with *planned buys* by undercover vehicles (Re: 1998 Los Angeles fraud cases). (2) Dispenser cabinets must be opened and examined to determine whether *skimmers* have been installed to copy payment card (credit, debit) information. This requires inspectors to collaborate with local police and credit card companies and to seize skimmers as evidence for court. (3) New methods will be required to regulate remote wireless calibration of parameters for retail motor-fuel dispensers and loading rack meters that bypass traditional wire-and-lead security seals on calibration mechanisms.

Development of New Devices. New weighing and measuring devices have been developed rapidly leaving NIST Handbook 44 technical codes lagging far behind: (1) Growing water shortages, primarily in the southern portion of the country prompted the installation of *water submeters* in multi-family buildings. This has led to the development of water submeter standards in NIST Handbook 44.[5] (2) Recently, a new tentative NIST Handbook 44 code has been adopted for transportation network company (e.g., Uber, Lyft) GPS measuring devices, or *virtual taximeters*, to charge fares. These GPS devices will eventually replace conventional electronic taximeters.[6] (3) Introduction of new *fuels* for motor vehicles to meet environmental pollution challenges and climate change require that new NIST Handbook 44 codes be developed. These fuels include: CNG, hydrogen and electric charging stations. (4) Wireless remote calibration of all electronic weighing and measuring devices has rendered obsolete traditional wire-and-lead security seals on calibration mechanisms and required the development of electronic audit trails as a stop gap measure. These have not proven very satisfactory since, experience has demonstrated, that all computer software can be 'hacked.'

In summary, many significant factors that affect projections are generally very difficult to anticipate or to quantify. As a result, most forecasting merely extrapolates current conditions into the near term.

In the Near Term

During the next ten years, it is likely that current trends in weights and measures regulation will continue but at an accelerated pace. In other words, unfortunately, problems providing effective regulation of weights and measures will probably worsen more quickly than solutions can be found.

Weighing and measuring devices. As mentioned previously, it is likely that the number of gasoline retail motor-fuel dispensers may peak soon at slightly more than three million because of the transition to alternative fuel vehicles announced by the automakers and mandated in proposed government climate change carbon reduction goals. Commercial scales appear to have already peaked at about one million. Traditional weights and measures practices of emphasizing device inspections over package inspections and 'inspecting every device every year' are obsolete as the retail sector of the economy transitions to a predominantly package-based marketplace. Over the past two decades, states have largely pushed weights and measures regulatory program budgets out of the General Fund to become self-funding with device registration fee revenue. However, this will not be viable in the long run. In the near term, device registration fees may be raised to offset the decline in commercial weights and measures devices but there will be increasing pushback from lobbyists arguing that it is too burdensome for a storefront retail industry already in decline.

Package inspection workload. Many weights and measures inspectors have limited experience with package inspection activities such as the complex *random sampling* procedures for checking the net contents of packaged goods or even the relatively simple *randomized sampling* procedures for verifying prices in retail store scanning systems. Few inspectors are familiar with 'method of sale,' 'labeling,' 'unit price code' or other package inspection activities because there are no registration fees or inspection fees to support these inspections as there are for device inspections. Moreover, weights and measures program managers have always prioritized device inspections over package inspections.

Unlike device owners, who bear complete responsibility for maintaining weighing and measuring device calibration within tolerance, retail stores, with few exceptions (e.g., meat departments at supermarkets), do not package the goods they sell. As a result, retail associations and lobbyists representing the retail industry often protest the conduct of package inspections at retail. However, package inspections targeted at production sites are not practical. Local and state jurisdictions, where large multinational production facilities are located, don't have adequate inspectors to monitor production lines filling millions of packages annually for national and international markets. Local weights and measures regulatory programs are limited to investigation of package complaints received from consumers, or other businesses, located in their legal jurisdictions. As a result, they must test '*net contents*' at retail or not at all. The NCWM opposes testing at retail because of small lots. Lobbyists for regulated industry oppose testing at retail because failed product lots are ordered off sale – an inconvenience for consumers – and the product must be returned to the distributor. Package inspections will continue to decline.

Weights and measures regulatory program budgets. The reduction in the number of weights and measures inspectors will most likely continue as the inventory of retail motor-fuel dispensers and commercial scales peak and begin to decline. The loss of device registration fee revenue will accelerate the reduction of inspectors. This situation will compel weights and measures regulatory programs to expand current efforts to prioritize device locations with poor compliance histories, extend the period between inspections, abbreviate inspection procedures from 'full inspections' to 'audits.' They must also sample devices at locations instead of inspecting all devices. It is likely that privatization of weights and measures device inspections will expand and the reduced inspector staff will be employed primarily in audits of the inspections by of private service companies. Weights and measures regulatory programs have two factors in their favor. First, they can survive, albeit down-sized, without general fund monies unlike most of government. Second, they perform services for the public with consumer protection inspections and complaint investigations. Since consumers are also tax payers and voters, this factor has often sheltered weights and measures programs from draconian (i.e., existential) budget reductions in the past. There will be a lot of pressure on weights and measures regulatory programs to find means to assess registration and inspection fees on package inspection activities. Some jurisdictions have implemented registration fees for price scanning systems but these will become commonplace in the near term.

Weights and measures inspectors. The number of weights and measures inspectors declined 25.9 percent from 1,835 in 2002 to 1,359 in 2019.[7] This trend is expected to continue in the near term as the number of weighing and measuring devices decline, budgets continue to shrink and privatization expands. All inspectors will need to be cross-trained to perform audits of inspections conducted by private service companies. In addition, new technology will replace existing devices with more advanced weighing and measuring devices so new hires will likely need a resume consisting of stronger math, science and computer skills. Periodic re-training will also be necessary to stay current. Inspectors will also need to become more efficient due to growing workloads so expensive specialized test equipment and vehicles will be required. For example, in Seattle, prover trucks with service boxes were introduced more than 25 years ago with built-in 5-gallon test measures (provers) for each grade of gasoline and integral storage tanks. However, in many jurisdictions, inspectors still hand carry and pour using two manual 5-gallon provers pushed on a welding cart transported in a pickup truck. The prover trucks reduce inspection time by approximately 50 percent and don't require inspectors to carry open provers and hand-pour the samples back into underground storage tanks for each grade of fuel dispenser. Timed hand-pours using large funnels unnecessarily expose inspectors to benzene vapors (e.g., leukemia hazard) and other harmful hydrocarbons.

In the Long Term

Looking out more than a decade, it seems possible that weights and measures regulatory programs may be nearly unrecognizable in comparison with the present. An extended example should serve to illustrate many of the challenges that face weights and measures inspectors in the long run.

*Electric Vehicle Charging Stations in California***.** The $2 trillion infrastructure bill, proposed by President Biden in April 2021, seeks funding to build 500,000 additional charging stations for electric vehicles (EV).[8] California is the leader in electric vehicles and charging stations as well as other alternative fuels (hydrogen, CNG). The state has 1,792 public access, privately-owned charging stations with 6,578 *Level 2* charging outlets and 338 charging stations with 3,375 *direct current fast charging* (DCFC) outlets.[9] Level 2 means, "*A medium charging speed (3.3–7.2 kilowatt), adding 12 to as much as 70 miles of range per hour.*"[10] DC fast charging means,

> The fastest charging currently available. DCFCs currently range from 50 kilowatt (kW) up to 350 KW, adding about 3 to 20 miles per minute, depending on the charger speed and state of charge of the battery. Most PHEV [plug-in hybrid electric vehicles], and some lower-range BEVs [100% battery electric vehicles] are not equipped with DCFC ports.[11]

Electric vehicle fuel costs are cheaper than gasoline.

> Drivers in California may expect to pay 30 cents per kWh to charge on Level 2, and 40 cents per kWh for DC fast charging. At these rates, the same Nissan LEAF with a 150-mile range and 40 kWh battery would cost about $12.00 to fully charge (from empty to full) using Level 2, and $16.00 with DC fast charging.[12]

The California Assembly passed Bill 808 in 2015. It included '*electricity*' in the definition of '*alternative fuels*' for motor vehicles. The new law was codified in the Business and Professions Code (BPC) at Division 5 "Weights and Measures," Chapter 14 "Fuels and Lubricants," Article 1, Section 13400(b)(4). This law enabled the California Division of Measurement Standards (DMS) to regulate privately-owned electric vehicle charging stations as 'measuring devices.'

> ***SEC. 3.***
> *Section 13400 is added to the Business and Professions Code, to read:*
> ***13400.***
>
> *For purposes of this chapter, the following terms mean the following:*
>
> *(a) "Advertising medium" includes banner, sign, placard, poster, streamer, and card.*
>
> *(b) "Alternative fuels" means:*
>
> *(1) "Biodiesel," a fuel comprised of mono-alkyl esters of long chain fatty acids derived from plant or animal matter that meets the requirements of the ASTM International Standard Specification D6751 "Standard Specification for Biodiesel Fuel Blend Stock (B100) for Middle Distillate Fuels."*
>
> *(2) "Biodiesel blend," a fuel comprised of biodiesel mixed with diesel fuel that meets the requirements of ASTM International Standard Specification D7467.*

(3) *"Dimethyl ether," an organic compound meant for combustion in compression-ignition engines that meets the requirements of dimethyl ether prescribed in this chapter.*

(4) *"Electricity," electrical energy transferred to or stored onboard an electric vehicle primarily for the purpose of propulsion.*

(5) *"Ethanol," denatured motor fuel ethanol meeting the requirements of ASTM International Standard Specification D4806.*

(6) *"Ethanol fuel blend," a motor vehicle fuel consisting primarily of ethanol mixed with gasoline meeting the standards prescribed for ethanol fuel blends by this chapter.*

(7) *"Hydrogen," a fuel consisting of high purity hydrogen intended for consumption in a motor vehicle with an internal combustion engine or fuel cell that meets the standards for hydrogen prescribed by this chapter.*

(8) *"Methanol fuel blend," a motor vehicle fuel consisting primarily of methanol mixed with gasoline meeting the standards prescribed by this chapter.*

(9) *"Natural gas," a gaseous mixture of hydrocarbon compounds consisting of primarily methane in the form of a compressed gas or a cryogenic liquid intended for use as a motor vehicle fuel.*

(10) *"Propane," a liquefied petroleum gas intended for use as a motor vehicle fuel and meeting the standards prescribed by this chapter.*

(11) *Any other fuel intended for use as a motor vehicle fuel that the secretary determines is an alternative fuel that has a standard specification from a standards development organization accredited by the American National Standards Institute (ANSI), or an interim standard specification pursuant to Section 13446.* [13]

County weights and measures regulatory bodies only have jurisdiction over commercial charging stations – those that charge a fee, *and* are not owned by a municipality. The publicly-owned charging stations are exempt from weights and measures regulation pursuant to California Attorney General's Opinion No. SO 77–13 dated November 22, 1977.

> The County and State Weights and Measures officials would not have jurisdiction over commercially used devices owned by a city.[14]

Subsequently, the California Division of Measurement Standards adopted regulations applicable to commercial electric vehicle supply equipment (EVSE) and effective January 1, 2021. The regulations modified the tentative device code, Section 3.40 "Electric Vehicle Fueling Systems," adopted by the National Conference on Weights and Measures and published by the National Institute of Standards and Technology in Handbook 44 *Specifications, Tolerances, and Other Technical Requirements for Weighing and Measuring Devices*. Detailed inspection procedures are contained in the DMS Examination Procedure Outline (EPO) No. 52 "Retail Electric Vehicle Fueling Systems." Specialized equipment, including a EVSE tester and an EVSE programmable load emulator, are required to conduct tests of charging stations. Presently, there is only one company that manufactures this test equipment and it is very expensive. The TS200 tests AC Level 2 charging stations (up to 75 amps). The TS400, which will replace it in 2021, is capable of testing both the AC and DC fast charge Level 2 charging stations (up to 8.25 KW).[15] DMS has two of the TS200 'loaners' for county weights and measures jurisdictions that haven't yet purchased test equipment.

At this time, county jurisdictions in California are registering commercial public access privately-owned charging stations and have not yet begun conducting tests. The registered charging stations will exclude 'existing' (in service before January 1, 2021) and publicly-owned charging stations which are, currently, a very large percentage of all commercial charging stations. County weights and measures programs will be required to investigate consumer complaints for existing charging stations but will not receive any fee revenue to support the testing. Moreover, publicly-owned commercial charging stations are exempt from inspections and registration fees but they compete directly with privately-owned charging stations.

The California metrology lab calibrates this test equipment. Electric vehicle supply equipment (i.e., charging stations) are required to indicate the electrical energy (megajoules MJ or kilowatt-hours KWh) unit price, and total price for each transaction. There must be a paper or electronic receipt including: total quantity of energy delivered, total computed price, unit price, maximum rate of energy transfer and type of current (e.g., 50 KW DC), any additional charges, total price of the transaction, unique EVSE identification number, business name and business location.[16]

One source of confusion for consumers is the method of sale. It is not possible for consumers to make value comparisons unless there is a standard unit price.

> At some stations, drivers pay for a connection or access fee; some pay for the kilowatts they received; some pay by how long they've charged. [17]

Typically, weights and measures jurisdictions adopt a device registration fee or inspection fee based on the actual expenses, including overhead, associated with inspection of a particular device. California has not established a specific registration fee for EVSE yet so the 'other' device category at BPC 12240(t) is applicable. County weights and measures jurisdictions may charge up to $20 for the device (EVSE), $100 for the location and the $2.20 state administrative fee.[18] Ultimately, the total annual fee revenue for EVSE must offset the decline in annual fee revenue for retail motor-fuel dispensers or the operating budget will be reduced with adverse consequences for inspector staff in the long term. Presently, there are 286,083 retail motor-fuel dispensers located in California.[19] The device registration fee for retail motor-fuel dispensers is $20 per device (BPC 12240(t)). In addition, there is a $2.20 per device administrative fee levied by DMS.

DMS records indicate that the inventory of *registered* alternative fuel devices in California include: hydrogen gas fuel dispensers (61) and compressed natural gas dispensers (479) and EVSE (108).[20] However, EVSE are expected to become the predominant alternative fuel device as automakers transition to electric vehicles. There are nearly 10,000 public access privately-owned charging station outlets at more than 2,000 locations but these are 'existing' devices and 'grandfathered' for ten years.[21] 'Existing' AC Level 2 devices (installed prior to January 1, 2021) and DC fast charge devices (installed prior to January 1, 2023) will be required to register after ten years – January 1, 2031 and January 1, 2033 respectively.[22]

Conclusions

In the long term, there is little question that weights and measures regulatory programs will be smaller: conducting fewer device and package inspections, employing sampling techniques instead of full inspections, abbreviating existing inspection procedures and expanding *privatization* of device inspections as well as package inspection activities. It is possible that businesses that package products will have their filling operations audited with video in lieu of time-consuming on-site net contents lot inspections. Check weighers, at the end of the production line, will probably continuously record weights for subsequent auditing by inspectors. Production oversight by management will need to be much more automated with alarms and automatic shutdowns in the event of short measure packages as they pass over in-line check weighers.

New, tentative, inspection procedures have been developed for testing by the National Conference on Weights and Measures, for new alternative fuels such as hydrogen and electricity. However, the National Type Evaluation Program (NTEP) is lagging well behind the California Type Evaluation Program (CTEP) testing and certification of devices developed for alternative fuels. In the future, more burden will be placed on industry for *self-regulation* including hiring private service companies to conduct inspections due to the unavailability of weights and measures inspectors. Weights and measures inspectors will spend most of their time auditing the work of private service companies and conducting investigations of consumer or producer complaints. Perhaps, a legal liability scheme will partially replace weights and measures regulation. Instead of making complaints to the weights and measures jurisdiction, consumers and producers would use small claims court or civil lawsuits to make damage claims against businesses that violate weights and measures laws.

It is likely that things will get worse before they get better. Elected officials generally are not known for being proactive. And, industry often opposes regulation until they experience disruptive events in the marketplace. Then, they tend to reluctantly embrace the relatively small expense of regulation by a disinterested third party like weights and measures officials. Too often, industry lobbyists will accept nominal regulation but oppose enforcement provisions with 'teeth in them' (enforcement provisions such as monetary penalties for noncompliance) or the lobbyists oppose adequate budgets to frustrate any effective field enforcement of weights and measures law. Facing smaller budgets, weights and measures regulatory programs will be compelled to 'downsize' their inspection and enforcement activities as described in the foregoing. This won't be enough.

Weights and measures program managers must re-imagine the entire process by starting with a blank slate. The responsibilities of businesses (sellers) and consumers (buyers) in marketplace transactions must be thoroughly re-examined. The transition from store transactions to internet transactions, accelerated by the COVID-19 pandemic; the transition from bulk sales to packaged goods; and the new technological innovations have all combined to render traditional inspections ineffective and inefficient. In a word – obsolete. Radically new techniques for verifying compliance with weights and measures laws must be invented. For example, compliance checks must be *designed into* all packaging and distribution processes rather than conducted at retail in response to a consumer complaint. This must necessarily require a new cooperative working arrangement among state weights and measures programs focused on production facilities. Businesses will probably have to hire independent auditor firms for weights and measures compliance just as they do with CPA firms to examine their financial records. Businesses will also likely have to purchase insurance to protect them from liability in weights and measures claims and lawsuits. The only thing that is certain about the future is that weights and measures regulation must adapt more promptly to try to keep up with changes in society, technology and the economy.

Notes

1 See Chapter 1, Table 1.7 "Total Weighing and Measuring Devices in the U.S."
2 Ibid.
3 National Institute of Standards and Technology. Handbook 44. *Specifications, Tolerances, and Other Technical Requirements for Weighing and Measuring Devices* (2020), Section 3.39 "Hydrogen Gas-Measuring Devices – Tentative Code."
4 NIST Handbook 44, Section 3.40 "Electric Vehicle Fueling Systems – Tentative Code."
5 NIST Handbook 44, Section 3.36 "Water Meters."
6 NIST Handbook 44, Section 5.60 "Transportation Network Measurement Systems – Tentative Code." For more information, see: Craig A. Leisy. *Transportation Network Companies and Taxis: The Case of Seattle*. New York: Routledge (2019).
7 See Chapter 1, Table 1.6 "Total State Weights and Measures Inspectors in the U.S."
8 The White House. "Fact Sheet: Biden Administration Advances Electric Vehicle Charging Infrastructure" *Statements and Releases* (April 22, 2021). www.whitehouse.gov/briefing-room/statements-releases/2021/04/biden-administration-advances-electric-vehicle-charging-infrastructure/ Accessed April 27, 2021.
9 U.S. Department of Energy, Energy Efficiency & Renewable Energy. "Alternative Fuels Data Center." www.afdc.energy.gov Accessed April 27, 2021.
10 California Governor's Office of Business and Economic Development. Tyson Eckerie and Gia Brazil Vacin (lead authors). "Electric Vehicle Charging Station Permitting Guidebook." (July 2019) 1st ed., p. 58. www.business.ca.gov Accessed April 27, 2021.
11 Ibid., p. 59.
12 California Air Resources Board. Drive Clean. "Electric Car Charging Overview" https://driveclean.ca.gov/electric-car-charging Accessed April 27, 2021.
13 California Assembly Bill 808 (2015).

14 State of California. *Attorney General's Opinions*, "Opinion No. 77–13 – November 22, 1977"
15 The TESCO TS200 Electric Vehicle Test System is suitable for AC Level 2 EVSE (up to 72 amps) but not DCFC charging stations. It has not yet been granted approval under the California Type Evaluation Program (CTEP) which certifies compliance with NIST Handbook 44. The TS400, which replace the TS200 in 2021, will test AC Level 2 and DC fast charge Level 23 (up to 8.25 KW). http://new.powermeasurements.com/products/ Accessed May 4, 2021. Approximate prices for the test equipment are: TS200 $26,000 and TS400 $52,000. Prices include support.
16 NIST Handbook 44, Section 3.40, op. cit., S.2.6 "EVSE Recorded Representations."
17 David Ferris and David Iaconangelo. "Industry fumes as Calif. takes measure of charging stations" (December 19, 2019). www.eenews.net/stories/1061847895 3. Accessed April 27, 2021.
18 California Business and Professions Code (BPC), Section 12240(t):

> (t) For all other commercial weighing or measuring devices not listed in subdivisions (g) to (r), inclusive, the device fee shall not exceed twenty dollars ($20) per device. For purposes of this subdivision, the total portion of the annual registration fee consisting of the business location fee and the device fees authorized by this subdivision shall not exceed the sum of one thousand dollars ($1,000), for each business location. A new regulation, effective April 28, 2021, increased the stater administrative fee from $1.10 to $2.20. The new fee schedule is published in Section 4075, Table A at www.cdfa.ca.gov/dms/pdfs/regulations/deviceadmin2021-0312-02SApproval.pdf .

19 California Division of Measurement Standards. *County Annual Report 2019–2020 Fiscal Year*. "Measuring Devices in Each County", p. 8 of 13.
20 Ibid.
21 U.S. Department of Energy. "Alternative Fuels Data Center", op. cit. (See Note 9).
22 California Division of Measurement Standards. "Electric Vehicle Supply Equipment Frequently Asked Questions." DMS Notice D-20-02. (June 16, 2020), p. 1 of 4.

Bibliography

California Division of Measurement Standards. "Electric Vehicle Supply Equipment Frequently Asked Questions." DMS Notice D-20–02. (June 16, 2020).

California Governor's Office of Business and Economic Development. Tyson Eckerie and Gia Brazil Vacin (lead authors). "Electric Vehicle Charging Station Permitting Guidebook." (July 2019) 1st ed.

National Institute of Standards and Technology. Handbook 44. *Specifications, Tolerances, and Other Technical Requirements for Weighing and Measuring Devices* (2020).

Glossary

See also:

(1) National Institute of Standards and Technology (NIST). Handbook 44. *Specifications, Tolerances and Other Technical Requirements for Weighing and Measuring Devices* (2020), Appendix D "Definitions."
(2) National Institute of Standards and Technology (NIST). Handbook 130. *Uniform Laws and Regulations in the Areas of Legal Metrology and Engine Fuel Quality* (2020), Section 1 "Uniform Weights and Measures Law," Section 1 "Definitions."
(3) National Institute of Standards and Technology (NIST). Handbook 133. *Checking the Net Contents of Packaged Goods* (2020), Appendix F "Glossary."
(4) International Organization of Legal Metrology (OIML), www.oiml.org

accurate device. A weighing or measuring device that conforms to the tolerances allowed by the applicable device code in NIST Handbook 44.

approval seal. Typically, an adhesive decal affixed to a weighing and measuring device by an inspector to prove that it passed inspection.

audit trail. Electronic calibration record of a weights and measures device. Needed to monitor changes to prevent fraud because remote electronic calibration easily bypasses wire-and-lead security seals intended to prevent tampering.

certificate of conformance (CC). Each weighing and measuring device used commercially must have a 'Certificate of Conformance' verifying compliance with the applicable technical code in the National Institute of Standards and Technology (NIST) *Handbook 44 Specifications, Tolerances, and Other Technical Requirements for Weighing and Measuring Devices* (current version). The CC is issued by the National Type Evaluation Program (NTEP) and administered by the National Conference on Weights and Measures (NCWM). Under NTEP, device prototypes are tested at designated state metrology (weights and measures) labs.

computing scale. A type of small scale that computes the charge for a commodity based on weight and unit price. Commonly, grocery stores use electronic retail computing scales with a 30-pound capacity. Price Look-Up (PLU) codes on the commodities (e.g., fresh produce) provide unit price.

Handbook 44. A NIST publication, updated annually, containing the technical standards for all weighing and measuring devices.

Joint Resolution of Congress (1836). Directed the Secretary of the Treasury to distribute sets of weights to customs houses and the states. These weights, an average of the customary system of units then in use, became the de facto standards for the United States: the avoirdupois 16-ounce *pound* (for goods sold by weight), a 128 fluid-ounce *gallon* measuring 231 cubic inches in volume (for goods sold by liquid measure), a *bushel* measuring 2,150.42 cubic inches in volume (for goods sold by dry measure), and a 36-inch *yard* (for goods sold by length).

labeling. Information indicating net contents and declaration of responsibility (manufacturer) on principal display panel.

legal for trade. Designation found on weighing devices meaning that the device is suitable for commercial use because it complies with Handbook 44 and it has a Certificate of Conformance (CC) number issued by NTEP.

legal metrology. According to the International Organization of Legal Metrology (OIML.org), the application of legal requirements to measurements and measuring instruments for the purpose of developing global standards.

liquid measure. Volumetric measures such as fluid ounces and gallons. When checking new contents of packaged goods that are liquids, inspectors use calibrated glassware designed for that purpose.

MAV. Maximum allowable variation or maximum error permitted in a sample while checking net contents of packaged goods.

method of sale. How a commodity is offered for sale (e.g., weight, volume, count, liquid measure, length). Solids are normally sold by weight and liquids are normally sold by volume.

model laws and regulations. Uniform laws and regulations developed and adopted by the National Conference on Weights and Measures and published by the National Institute of Standards and Technology in Handbook 130.

moisture loss. Reasonable variation from net contents declaration on the label of a packaged good due to moisture loss during distribution. This was a departure from the traditional requirement that the actual net contents of a packaged good equal labeled net contents at the time of sale.

National Conference on Weights and Measures (NCWM). A national standard-setting organization, whose members are weights and measures officials – both state directors and individual inspectors – from all of the states and territories. Industry representatives are associate members who cannot vote on changes to weights and measures model law or technical codes but may participate in discussions by standing committees.

National Institute of Standards and Technology (NIST). An agency of the U.S. Department of Commerce that includes the Office of Weights and Measures (OWM). OWM seeks to negotiate uniform weights and measures laws with other nations and promotes uniform laws, inspection procedures and technical standards among the states and territories. It is not an enforcement agency. NIST is located in Gaithersburg, Maryland.

NTEP (National Type Evaluation Program). A program supervised by the National Conference on Weights and Measures (NCWM) wherein specific state metrology laboratories conduct tests to determine whether new weighing and measuring devices meet all the requirements of the relevant national technical standards contained in NIST Handbook 44.

net weight. The gross weight less the tare weight. Weight of the commodity only.

origins of weights and measures in antiquity. The earliest weighing and measuring double-pan balances and polished stone weights were used by the river civilizations of the Indus River, Euphrates River and Nile River approximately 5,000 years ago.

package inspection activities. Enforcement of weights and measures laws pertaining to goods offered for sale in packaged form: e.g., net contents of packaged goods, price scanning systems, unit price code, method of sale, package labeling.

packaged goods. Commodities that are sold in packaged form rather than from bulk across a scale or through a meter.

point-of-sale. A direct sale where buyer and seller are both present. Checkout lanes at supermarkets have POS systems to weigh commodities sold from bulk (e.g., fresh produce) and compute price.

price scanning system inspections. Verifying that advertised prices and shelf price tags are the same as the scanned price (bar code) in the store computer at checkout.

random pack. Packages of a commodity with varying net contents.

ready-to-eat foods. Restaurant-style food for immediate consumption without further preparation. Sold as a meal and not by net contents.

standard pack. Packages of a commodity offered for sale with identical net contents.

tare. The weight of all packaging materials. Everything but the commodity itself.

tolerances. Allowable errors of package net contents under-registration or over-registration (i.e., short weight or short measure). Over-registration occurs when a commercial scale or meter overstates the actual amount of product delivered. This would be a minus error during an inspection because less product is delivered than indicated on the device display.

unused dry tare. Weight of new (unused) packaging materials. Normally the gross weight less the net weight of a good is tare.

unit price. Price per unit of measure for a packaged product (e.g., price per fluid ounce), or a commodity sold from bulk. Unit price is normally shown in advertisements and displayed on retail store shelf tags in order to allow the consumer to make value comparisons. Unit price serves as a common denominator among different size packages.

weights and measures. A system for determining the quantity of a commodity sold in commerce. Usually refers to weighing and measuring devices. As consumers, nearly everything we purchase is sold by 'weight' (ounces, pounds) or 'measure' (e.g., liquids: fluid ounce, gallon; or solids: dry pint, bushel).

wet hose. Metering system where the discharge hose is full of product to the nozzle at the start and end of delivering product. For example, retail motor-fuel dispensers (gas pumps).

wet tare. Weight of used tare. Includes free fluids that may have been absorbed by soaker pads.

Appendix A: Comparison of 2002 and 2019 Surveys of State Weights and Measures Regulatory Programs

Inspectors and Devices

Metric	*2002*	*2019*	*Change*
W&M Inspector Ratios[a]			
W&M Inspectors	1,835	1,359	(25.9%)
Budget/Inspector	$73,.853	$120,480	63.1%
Population/Inspector	157,149	241,011	53.4%
Land Area/Inspector	1,928 mi^2	2,774 mi^2	43.9%
RMFD/Inspector	1,377	2,174	57.9%
Scales/Inspector	550	749	36.2%
W&M Devices[b]			
Retail Motor Fuel Dispensers	2,526,343	2,953,843	16.9%
Commercial Scales (all kinds)	1,010,155	1,017,217	0.7%
Total	3,536,498	3,971,060	12.3%

Notes

[a] 2002 State Survey: Projected from 39 states/86.5% of U.S. population (50 states). 2019 State Survey: Inspectors projected from 26 states/68.6% of U.S. population (50 states); budget projected from 17 states/34.2% of U.S. population; Retail Motor-Fuel Dispensers (RMFD) projected from 28 states/72.4% of U.S. population; scales projected from 24 states/69.5% of U.S. population. Projections assume that metrics are proportional to population. In a few instances, extreme outlier data provided by a state was not included in a projection to avoid skewing it. See Table 1.6 "Total State Weights and Measures Inspectors in the U.S."

[b] 2002 State Survey: RMFD projected from 37 states/81.8% of U.S. population; scales projected from 37 states/ 82.3% of U.S. population. 2019 State Survey: RMFD projected from 28 states/72.4% of U.S. population; scales projected from, 24 states/69.5% of U.S. population. Projections assume that metrics are proportional to population.

Appendix B: Comparison of 2002 and 2019 Surveys of State Weights and Measures Regulatory Programs

Inspections, Frequency, and Failure Rates

Metric	*2002*	*2019*	*Change*
RMFD Device Inspections[a]			
RMFD Inventory	2,526,343	2,953,843	16.9%
Percent Inspected	83.4%	87.4%	
Frequency of Inspection	1.2 years	1.1 years	
Percent Failed	6.6%	6.9%	
Scale Device Inspections[b]			
Scale Inventory	1,010,155	1,017,217	0.7%
Percent Inspected	87.5%	N.A.	
Frequency of Inspection	1.1 years	N.A.	
Percent Failed	6.9%	N.A.	
Package Inspections[c]			
Price Scanning	62.5% of states	54.5% of states	
Net Contents	90.0% of states	63.6% of states	

Notes

[a] 2002 State Survey: RMFD projected from 39 states. 2019 State Survey: RMFD projected from 28 states/72.4% of U.S. population. Projections assume W&M devices are proportional to population. See Table 1.8 "W&M Device Inspection Frequency and Failure Rate."

[b] 2002 State Survey: Scales projected from 40 states. 2019 State Survey: Scales – no data available due to incomplete survey information. Projections assume W&M devices are proportional to population. See Table 1.8 "W&M Device Inspection Frequency and Failure Rate."

[c] See Table 1.9 "State Package Inspection Activities."

Index

For Product Safety Concerns and Information please contact our EU representative GPSR@taylorandfrancis.com
Taylor & Francis Verlag GmbH, Kaufingerstraße 24, 80331 München, Germany

www.ingramcontent.com/pod-product-compliance
Lightning Source LLC
Chambersburg PA
CBHW061318220326
41599CB00026B/4943

* 9 7 8 1 0 3 2 2 0 4 6 5 9 *